Forest Project Analysis and Management

Forest Project Analysis and Management

Ravinder Dhanai

CYBER TECH PUBLICATIONS
4264/3, Ansari Road, Daryaganj, New Delhi-110002 (India)
Ph.: 011-23244078. Fax: 011-23280028
E-Mail: cyberpublicationsdelhi@yahoo.com

Forest Project Analysis and Management
Ravinder Dhanai

First Edition 2009

Published by :

G.S. Rawat for **Cyber Tech Publications**
4264/3, Ansari Road, Daryaganj, New Delhi-110002 (India)
Ph.: 011-23244078. Fax: 011-23280028
E-Mail: cyberpublicationsdelhi@yahoo.com

Printed at : **Amit Enterprises**, Maujpur, Delhi, (9810878184)

Preface

Progress has been made when projects are planned and implemented in collaboration with local communities. However, communities vary in their interest and capacity to manage their tree and forest resources. Why have some communities planted or managed their trees and forests well for generations while others have not? Why have some working management systems recently broken down but others have continued? Often investigation shows that when forest management is successful there are sets of rules and expectations related to tree use and management that the community respects. Where incentives for the community to follow these rules have been weakened, management often breaks down. In order to stimulate increased community efforts to improve and manage trees and forests, forestry activities must recognize and support local rules and expectations and provide incentives for local investment in tree and forest resources. There is a need to develop institutional frameworks for all those involved in order to have a better understanding of how to better support local institutions in forest management.

Forest closely links human activities with nature. Similar to agriculture, forests also reveal the wealth of a nation. Investment in forestry and realisation of returns is taken care of by the state, due to the long gestation period and several laws governing the exploitation of forest resources. However, efforts are taken by the government to widen the forest cover through several afforestation programmes and agroforestry projects. Both at the regional level and at the micro level, the stakeholders in the forest development have to understand certain basic principles underlying forest management. An attempt is made in this book to explain the principles of economics and management as applied to a development project in the forest sector. This book is prepared with the undergraduate students of forestry in mind and hence the sophisticated mathematical applications are limited. It is believed that this book will be useful to them.

I hope that this research work of mine would be informative to the readers and they will derive much pleasure going through it.

—***Author***

Contents

1

World Bank's Forest Policy Implementation Review

The World Bank is conducting a review of how its 1991 Forest Policy is being implemented in practice. At the same time, it is developing a new strategy to guide its future investments in forestry. It would be far more logical if the Forest Policy Implementation Review would precede the development of a new strategy in order to ensure that the results of the review could be absorbed, debated and turned into lessons learned. The lessons drawn from the Forest Policy Implementation Review should be at the core of the development of a strategy for Bank activities that are related to forests since the key lies in implementation and not in policy or strategy development.

The first section of this paper explains why World Bank Forest Policy matters deserve continuing attention on the part of the environmental, social justice and indigenous peoples movements. The next sections provide a short history of the World Bank's 1991 Forest Policy and an overview of what is new since the adoption of that policy. This is followed by an explanation of the FPIRS process as well as related processes in which the World Bank is involved. The next section describes key NGO-concerns and is followed by a set of specific recommendations.

IMPORTANCE OF THE WORLD BANK'S FOREST POLICY

- The World Bank is the single largest source of 'development' finance and of investments in the forest sector globally

- The World Bank Forest Policy is very influential and serves as a model for other multilateral development banks and bilateral aid agencies
- The WB Forest Policy covers all types of World Bank activities

(1) It applies to the entire World Bank group. This means that in addition to the Bank's regular lending activities, the policy applies to its lending to the private sector through the International Finance Corporation (IFC) and to the granting of financial guarantees through its Multilateral Investment Guarantee Agency (MIGA);

(2) It cover all sectors and activities financed by the World Bank. Given its intersectoral nature, the policy does not only cover investments in forestry or natural resource management, but also infrastructure development, transport, mining, dams, as well as the impacts of macro-economic policy prescriptions which lie at the core of structural and sectoral adjustment programs.

(3) The WB Forest Policy increasingly covers private sector investments, wherever there is World Bank co-financing. An example is the $ 3.5 billion Chad/ Cameroon Oil and Pipeline project promoted by three of the world's largest oil companies, Exxon, Shell and ELF. The World Bank group's planned co-financing of the project is less than 10% of total project costs, but since the pipeline will traverse largely intact rainforest areas which are home to indigenous people, the Forest Policy must be applied both with regards to the direct impacts (pipeline construction) as well as to the induced impacts (increased access of loggers, poachers and migration).

A SHORT HISTORY

Controversy about forest destruction has been the Bank's Achilles Heel for nearly two decades. The flagship disasters in

Brazil and Indonesia in the 1980s focused on the role of the Bank as forest destroyer. The subsequent attempts by the Bank to do good forestry through the TFAP (Tropical Forest Action Plan) proved a further public relations disaster for the Bank. The TFAP was an ambitious plan launched by the World Bank and other international actors in the mid-1980s to mobilize US $ 8 billion over a period of five years. While the official goal was to curb deforestation, the TFAP's planned investments were very much geared towards industrial logging activities. It was the uniting of ecological justice NGOs in opposition to the TFAP that created the WRM. The TFAP debacle, which the WRM triggered, and broad-based criticism of environmentally destructive forestry loans for West Africa forced a rethink of forest policy in the Bank in the late 1980s and early 1990s. The WRM network was influential in this policy review, which resulted in a new World Bank Forest Policy in 1991.

The 1991 Policy was an important step forward for the Bank. It committed the Bank to a new approach to forests including commitments to:

- assess the likely impacts on forests of macro-economic and non-forest sector loans: SAPs, agricultural projects, dams, mines etc.
- take into account the interests of communities in establishing plantations
- involve forest-dwelling people in establishing protected areas
- avoid funding logging in primary moist tropical forests
- condition other lending for forest operations on the inclusion of measures to safeguard the interests of forest-dwellers
- experiment with new projects with forest-dwellers including in promoting non-wood products.

In 1994 the Bank carried out a Forest Policy Implementation Review, partly as a result of insistence of NGOs and a supportive US Executive Director. This was a rather unsuccessful stock-taking exercise, although the scope given for NGO participation in the process was a step forward in Bank procedures. A half-hearted attempt by some Bank staff to have the proscription on funding of logging in primary moist tropical forests was lifted by sustained NGO pressure. The WRM was especially critical of the failure of the review to look at:

- the cross-sectoral impacts of Bank lending on forests
- the lack of staff incentives to observe the policy, especially with respect to adequate participation
- the social impacts of Bank lending related to forests, especially in terms of impacts on women, indigenous peoples and land tenure, or measures to address these impacts (see annex 2).

The Bank admitted that these were real deficiencies in its review but that it was obliged to report to the World Bank's Board of Executive Directors (EDs) that year and so had no time to fill the gaps. The WRM heavily lobbied the EDs at the Bank not to accept the Review as it stood. In the end they did accept it on condition that a follow up review was carried out in 1997. After many delays this revised review is now underway.

What's New?

Since 1991, there have been a number of new developments which make this a significant moment for the Bank and for forests:

- The Bank is now active in transition countries and is starting forestry projects there: the 1991 policy was essentially focused on tropical forests only. This might result in greater pressure on forests and processes similar to those occurring in the South.
- Bank lending is increasingly through programme lending and is less project linked.
- The Bank seeks to direct more funding through the International Finance Corporation (IFC), its branch which lend directly to the private sector. Other support of private sector investment is channeled through the World Bank's

MIGA (Multilateral Investment Guarantee Agency, which provides political risk insurance to private businesses.

- The Bank is seeking to become the financial channel for forestry funding linked to the Kyoto Protocol (Clean Development Mechanism), which might result in massive substitution of natural forests by monoculture tree plantations.
- Global negotiations on forests are nearing consensus on some issues which we believe to be less relevant and which might form the basis for a forests convention. The Bank is seeking a role in implementation of any international agreement which arises.The issues on the table at the global negotiations are questions related to 'national sovereignty', 'sustainable forest management', 'criteria and indicators for sustainable forest management', 'National Forest Programmes', 'Promotion of Trade in forest products', 'voluntary self-regulation' etc. In contrast, the issues we want to be addressed are indigenous peoples rights, participatory agrarian reform, community control of forests, reforms in aid, regulation of transnational investments, limits on cross-sectoral forces pressing on forests, halt in illegally produced forest products, the equivalence of international laws on human rights and the environment with international laws on trade.
- A new, feisty set of World Bank forestry employees are much more aware of the role of forestry in national political economies but remain convinced that, done right, forestry can be socially and environmentally sustainable and good for national development. They are seeking to use the heavy conditionality of structural and sectoral adjustment lending to reform developing country forestry policies and practices. Critical facts are simply being ignored such as the lack of evidence that large-scale timber extraction can be sustainable, meet local people's needs or contribute meaningfully to poverty alleviation, while its political effects in concentrating power in the hands of often corrupt elites is a major obstacle to national development.

SECOND GUESSING THE BANK'S VISION

There is still a great deal of debate within the Bank about the appropriate way forward when it comes to forestry. A vision can be seen to be emerging, however.

This vision seems to include the following elements:

- Tropical forests will be all but gone in 50 years if current trends continue
- Staff in developing country forestry departments are becoming less not more adequately trained.
- Key reforms are needed to promote collaborative forest management and reformed forest governance.
- Curbing forest piracy is essential.
- Sustainable Forest Management (at least sustaining flows of some key goods and services) is possible if there is a level playing field in terms of standards, regulations and costs.
- The Bank should therefore support a Global Forests Convention.
- Rates of return on SFM are however not competitive with the return on capital expected by private investors today, especially if the costs of protection and provision of ecological services are included.
- State regulation of forestry and laws to maintain PFE (Permanent Forest Estate) are thus crucial.
- Collaborative forestry options are very useful but not a panacea.
- Logging is not really responsible for forest loss.
- Certification should be promoted.
- The prohibition on Bank funding of logging in primary moist tropical forests should be lifted.
- The Bank should integrate forestry issues into Country Assistance Strategies and into structural adjustment conditionalities.

The FPIRS Process

The process of the current FPIRS review has already been substantially changed owing to NGO lobbying. An initial proposal circulated at the very end of 1997 was to carry out a hurried implementation review and strategic planning process in the early months of 1998. This was dropped when senior Bank management agreed with NGOs that it was hasty and ill-thought out.

As conceived now an 'independent' FPIR will be carried out by the Operations Evaluation Department during 1999. Regional consultations are being planned for the second half of 1999 and will be complemented by country studies. A draft OED report is to be discussed by an 'Advisory Group' in the summer of 1999. The final report is scheduled to go to the World Bank's Board of Executive Directors later in the year. Jointly with other NGOs, the WRM submitted comments on the OED process 'design paper' to the Bank in early 1999.

Simultaneously, Bank staff from the Environment and Rural Development Departments will undertake a Stategy Development (SD) process, which will try to ascertain what the Bank's policy should be for the coming millennium. The results of the OED review are meant to feed into the SD which is meant to come to conclusions by April 2000. Details of the process can be found on the FPIRS website (see below for details). The Bank is contracting a number of agencies and individuals to write 15 'Analytic Papers' which are to be used to help stimulate discussions in the regional consultations and in the discussions of the draft Strategy paper. These are encouragingly broad and should stimulate a broad debate about the future of Bank involvement in forest-related policy and lending.

Participation: The Bank had initially contracted the IUCN to mediate the Bank's interactions with NGOs. However, NGOs including the WRM, objected strongly to this. As a result, the Bank's own NGO liaison unit will now help with NGO involvement but the IUCN is still being retained 'to support the Bank in developing and implementing a review process that remains transparent and open to all interested parties', though this has not until now resulted in positive feedback to and from NGOs.

Related Processes

Meanwhile the Bank is involved in two other related processes as a result of Wolfensohn's commitment to do something about forests. It has struck up an Alliance with the WorldWide Fund for Nature to promote the certification of logging and it has adopted WWF targets for setting up protected areas to cover 10% of the major forest types in all borrower countries.

This involves close collaboration with the WWF in forest policy discussions with WWF staff seconded to the Bank for liaison purposes.

The Bank is also engaged in a series of meetings with the CEOs of major forest industry companies to discuss best practice and see how industry can move towards this. Some NGOs and the IUCN have been involved in this process, but the early meetings conspicuously avoided addressing the social and political concerns. The WRM has written to these NGOs and the Bank to highlight the need to address social issues in discussions with industry and we have made recommendations that social NGOs, Indigenous Peoples Organisations and Trades Unions should be involved in future meetings.

KEY CONCERNS

OED Process

Participatory field studies of actual projects will not be carried out.

The procedure for selecting NGOs for the regional meetings is not transparent.

The OED draft will only be seen by the Advisory Group. The process should be more open and interactive.

It is not clear that the OED review will look at the IFC and MIGA. Mark Egstein of the World Bank's OEG is meant to be carrying out a study of IFC activities but nothing is planned regarding MIGA.

Environment department staff at the Bank are known to have little confidence that the OED review will look at the social issues.

Strategy Development Process

The Strategy Development process is being carried out simultaneously to the OED study when logically the review should precede it. There is a risk that Bank views will become entrenched and justified in the strategy development before the OED study can feed into it.

The degree to which the strategy development process will be open and participatory is not clear. Debate on early drafts and even later drafts does not seem to feature. In this sense the process seems less open than in 1994.

Plantations

Although tree plantations were not a major issue in the 1991 policy they had assumed far greater significance by 1994. The final review advocated much more lending for tree plantations under the guise of 'Resource Expansion'. The Bank's enthusiasm for CDM and joint implementation, related to the Kyoto Protocol, means that there is likely to be heavy emphasis on plantations in the strategy development even when there is growing evidence that large scale plantations result in highly negative social and environmental impacts, which can include forest clearing.

Recent Bank structural adjustment lending in South East Asia has also been aimed at expanding oil palm plantations. Tree crop estates are increasingly being developed by the private sector, creating resource conflicts with local communities and leading to increasing deforestation.

Forestry Projects

Bank staff are keen to lift the current prohibition on the Bank funding logging in primary moist tropical forests, whereas many NGOs want to see this prohibition extended to other forest types especially the remaining fragments of old growth, tropical dry, boreal and temperate forests.

Forestry is treated as a strategic sector of importance to climate, biodiversity and national development objectives. Local, social and poverty alleviation aspects take second place in these discussions.

Community Forestry and Collaborative Protected Area Management

While the Bank seems interested in funding community forestry and collaborative protected area projects, the Bank does not seem to emphasise the importance of tenure reforms, the special concerns of women and indigenous peoples.

The Bank may use the argument that it should fund community-based forest management as a justification for lifting the ban on funding logging in primary moist tropical forests.

Cross-Sectoral Issues

The most worrying issue remains the forestry focus of the Bank's approach. It seems that the World Bank's Vice Presidency for Environmentally and Socially Sustainable Development can't cope with the idea of having to reform the other sectors of the World Bank, although this is clearly necessary to address cross-sectoral causes of forest degradation and deforestation.

Recommendations to Influence the Process:

- Inform your country's Executive Directors as soon as possible of your concerns about the FPIRS
- Demand that the OED carried out participatory reviews of specific projects with affected communities to ensure that a real assessment of implementation is achieved
- Ensure that the FPIR looks at all sectors of Bank lending that affect forests and reviews the performance of all members of the Bank group including MIGA and the IFC
- Insist that the Strategy Development process be re-scheduled to follow after the completion of the FPIR
- Query the modalities for NGO and CBO participation in the process and demand clarity on the procedures that will be used to identify participants for regional and other consultations.

Recommendations to Influence the Bank's Policy:

- Create internal institutional mechanisms and oversight (p.e. incentives and penalties) to ensure World Bank staff compliance with the Forest Policy in all lending acitivities (projects, sectoral and structural adjustment programmes) as well as non-lending activities (Country-Assistance-Strategies, Technical Assistance Programmes).

- Develop the Bank's capacity for informed and effective participation by NGO and community based organisations in forest related projects and policy debates
- Address underlying causes of deforestation by addressing questions of indigenous rights, land security, agricultural stabilization, alternative income generating activities, and the impacts of other sectors including especially the energy, transport and mining sectors.
- Expand the ban on direct financing of logging in moist primary tropical forests to dry tropical forests, temperate and boreal forests.
- Ban funding of projects that damage or destroy any of the world's remaining primary forests, including logging operations, roads, mines, pipelines, tree plantations, etc..
- Suspend all loans to large-scale tree plantations until a specific and participatory assessment of the impacts of such plantations on forests, people and the environment is carried out. All types of tree plantations must be subject to this type of assessment, including plantations for pulpwood, timber, palm oil and carbon sinks.
- Establish a permanent mechanism to monitor and evaluate the World Bank's forest-related activities.

CROSS SECTORAL COMMITMENTS

- 'The Bank will give closer attention to infra-structural and other land-using projects to minimize their potentially negative effects on forests.' (Sector:62).
- 'the Bank's economic and sector work should address forestry issues within a multisectoral context...(including) reforms of concession policies, forest revenue systems, fiscal, tax, and agricultural policies, infrastructural and

land use planning procedures, land tenure systems and enforcement mechanisms' (Sector:63).

- 'forest sector policies advocated during project preparation will grow out of those put forward in the country economic memorandum, the national Environmental Action Plan, and the Consultative Group process... and as structural or sectoral adjustment inputs' (GBP:art 4). The guidelines advocate paying special attention to macroeconomic and policy issues such as revenue sharing, resource pricing, public investments in other sectors, institutional budgets, tarriffs, energy policy, incentive structures relating to land use (GBP:art 5)
- 'Financing of infrastructural projects (such as roads, dams and mines) that may lead to loss of tropical moist forest and other primary forests will be subject to rigorous environmental assessment as mandated by the Bank's operational guidelines for projects that raise diverse and significant environmental issues. A careful asssessment of the social issues will also be required' (Sector:65).
- 'In countries... in which forests are seriously threatened by encroachment and destruction... special efforts will be made to support economic development in poor, densely populated areas around the forests or in the origin areas of forest encroachers' (Sector:65).

Tree Crops

- 'The Bank will increase its efforts to finance the creation of additional forest resources (sic.)... in potentially productive degraded forests, wastelands, forest fallows, shrublands and abandoned farmlands. The interests of communities that depend on such areas will have to be considered on setting target areas' (Sector:64).

Protected Areas

- 'The Bank will support initiatives to expand forest areas allocated as parks and reserves to institute effective management and enforcement in new and existing areas...

The Bank will stress new approaches to management of protected areas that incorporate local people into protection benefit sharing and planning and will highlight the need to consider the needs and welfare of forest-dwelling people' (Sector:65).

Logging

- 'The Bank Group will not under any circumstances finance commercial logging in primary tropical moist forests' (Sector:65)
- Except for projects for forest protection and those oriented toward small farmers, 'other lending operations in the forest sector will be conditional on government commitment to sustainable and conservation-oriented forestry (which)... entails': sound policies, a plan, env. & soc. assessments, adequate institutional capacity and 'setting aside adequate compensatory forests to maintain biodiversity and safeguard the interests of forest dwellers, specifically their rights of access to designated forest areas' (Sector:66).
- 'The Bank does not finance commercial logging operations or the purchase of logging equipment for use in primary tropical moist forest. In borrowing countries where logging is being done in such forests, the Bank seeks the government's commitment *to move toward* sustainable forest management of those forests... and to retain as much forest cover as possible. Where the government has made this commitment, the bank may finance improvements in the planning, monitoring, and field control of forestry operations to maximise the capability of responsible agencies to carry out the sustainable management of the resource' (OP:art 1a). These broad policy changing projects should take 'into account limits on absorptive capacity and management control' (GBP: art 8).

Forest Dwellers

- 'Experimental programs to test alternative approaches to the participation of local people and to the promotion of nonwood products of natural forest to benefit such people will also be financed' (Sector:65).

- Financing of forestry projects will depend on the borrower government adopting a 'comprehensive and environmentally sound forestry conservation and development plan that clearly defines the roles and rights of government, the private sector and local people (including forest dwellers).

SUMMARY

The World Bank's 1991 Forest Policy marked an important step forward in attempts by the World Bank to limit the negative social and environmental impacts of its lending. It gave a priority to curbing deforestation, popular participation and poverty alleviation and sought to introduce across-the-board measures into World Bank operations to realise these objectives as they related to forests.

The 1994 'Forest Policy Implementation Review' was meant to be an attempt to appraise the extent to which the commitments made in 1991 have been adhered to in Bank operations since that date and to what extent this had been effective and appropriate.

Problems with the Review

Unfortunately, as is now clear from the April draft of the 'review', there are some worrying omissions and limitations to the review.

1. Lack of Attention to Operational Incentives

The review has paid inadequate attention to the operational structures of the Bank. While it notes that some important commitments that were made in the policy have not been taken into account in many Bank forestry projects since 1991, it has made no attempt to identify the extent to which this results from the lack of incentives for Bank operational staff to abide by social and environmental policies. This is a serious omission which must be corrected if useful conclusions are to be drawn from the review about how to improve Bank performance.

2. No Consideration of Operational Policy Notes

Although the Forest Policy was issued in 1991, it was not until 1993 that an Operational Policy Note (4.36) and notes on Guidelines and Best Practice (4.36) were issued to Bank staff. The review makes no attempt to examine whether these operational notes adequately encapsulate the commitments of the policy (in our view they do not), why they were issued so late and whether this affected the implementation of the policy.

3. Failure to Examine Cross Sectoral Issues

The 1991 document was a 'Forest Policy' which was expressly meant to apply to Bank lending as a whole and not just forestry and natural resource management projects, which constitute only 5% of Bank lending.

The review does not seek to assess the extent to which Bank lending for structural or sectoral adjustment since 1991 has adhered to the new forest policy nor assess the extent to which the policy was applied in other sectors, notably agriculture. This is a serious omission which it will take time to remedy. The main impacts of Bank lending on forests are indirect, resulting from non-forest sector lending, and the Bank must take stock of these impacts if the review is to make any sense.

4. Anecdotal, Unsystematic, Literature-based Approach

The review is methodologically flawed. It fails to identify clear indicators by which to evaluate project performance, to see if they have adhered to the objectives 1991 Forest Policy and other Bank goals. It fails to elucidate whether project documentation, supervision mission reports or project completion reports provide adequate data for a proper review. No field studies-not even sample studies-were undertaken to check the extent to which the information available to the review team corresponds to reality.

The review perpetuates the simplistic notion, repudiated by previous OED studies by the Bank itself, that project performance can be measured in terms of the time spent on supervision or sector work. The result is an anecdotal and inconclusive review.

5. Misleading Evaluation of Participation

Considering the high priority given to a more participatory approach, the analysis of the quality of participation in recent Bank lending is disappointingly shallow.

No comparison is made of the differences between NGO and Bank perceptions of their involvement in Bank projects. No attention is paid to the problems local and national NGOs continue to experience in getting timely information in the right languages.

The review does highlight the need for effective participation in decision-making by other 'stakeholders' in order to press through policy reforms that are resisted by borrower government agencies and political elites. Given this, it is crucial that the review makes a detailed evaluation of the extent to which the Bank has been able to achieve this goal. This will require extensive consultation with NGOs.

6. The Bank's Political Role not Questioned

The recognition by the Bank that effective forest policy reform in borrower countries requires the 'empowerment' of other sectors of civil society ('stakeholders') to challenge vested interests raises the question of whether effective Bank lending related to forests is necessarily political in nature and thus in violation of the Bank's Articles of Agreement.

At the NGO Consultation in London, Bank staff admitted that in practice the Bank is in violation of its Charter both in project preparation and implementation.

The review needs to confront this contradiction and examine the extent to which its structural relations to borrower countries, as an intergovernmental development assistance agency, mean that its current lending necessarily reinforces the position of these vested interests.

7. Lack of Attention to Poverty Alleviation and Women

Despite the fact that poverty alleviation is an express priority of the Bank's and one highlighted in the Forest Policy, the review pays very little attention to this issue. Almost no mention whatsoever is made of the role of women in the Bank's forest-related lending since 1991. These are grave deficiencies, which must be remedied if the review is to effectively contribute to improved Bank lending.

8. Poor Analysis of Land Rights Issues

The review recognises that tenure reforms are crucial to sound forest management and to ensure that local communities' interests are secure in the long term. However, the review's analysis of this important issue is both very limited and analytically flawed. The review fails to distinguish between tenure rights, conditional community management leases and joint forest management arrangements. It also fails to examine the large number of current Bank projects-notably the Natural Resource Management Projects in Brazil-where land conflicts have intensified owing to inadequate attention being paid in project preparation to national and state land policies.

9. Weak Treatment of Indigenous Peoples

Although the review reveals that Bank projects since 1991 have been singularly unsuccessful in dealing with indigenous peoples, the review does not identify the reasons for this failure in implementation. If the review is to contribute to resolving this problem, it needs to look at the many projects which failed to deal with indigenous peoples and discover why the issue was ignored.

The 1991 Forest Policy makes a commitment not to fund projects in borrower countries where forest policies, explicitly including policies towards indigenous forest dwellers, are at odds with the Bank's. However, the review fails to examine why Bank forestry projects continue to be funded where national policies towards forest dwellers are directly contrary to the Bank's (as they are for example in Indonesia and Thailand).

10. Ambiguous Position on Primary Forest Logging

The 1991 Forest Policy took the important step forward of expressly prohibiting Bank funding of logging in primary tropical moist forests. However, the review fails to examine if the Bank has kept this commitment. Instead, the review clearly advocates increased Bank support for 'harvesting operations' and claims that 'natural forest harvesting' (ie logging) can be an alternative to deforestation, despite the evidence that 'harvesting' on the scale needed to qualify for World Bank funding accelerates deforestation.

In view of the dramatic destruction of the world's forests, it is discouraging to see the institution with the world's largest portfolio affecting forests creating a spurious distinction between 'financing logging' and 'supporting harvesting'. While-as the review recognises-logging is such a profitable business it does not require direct financial support, the Bank provides and plans to increase indirect subsidies for logging in a variety of ways, by providing management plans, inventories etc..

The ban on financing logging must go beyond a public relations device. As long as all existing evidence points to logging as a major cause of deforestation, the Bank should not subsidize it in any shape or form, but encourage alternative income-generating activities, land security and agricultural stabilization.

11. Weak Treatment of Temperate and other Forest Types

Since the 1991 Forest Policy was adopted, political and economic changes in Eastern Europe and the former Soviet Union have led the Bank to open its lending to this region, with major new implications for forests. Already the Bank is engaged in several new projects in these countries, many of which are quite controversial.

The review's treatment of this crucial new area is very weak and suggests that rather than learning from its experiences in tropical countries, the Bank is about to repeat the past mistakes-over-enthusiastically promoting increased logging without due regard for the social and environmental impacts or taking adequate account of the political and institutional limitations of borrower governments. Much more detailed studies and consultation with NGOs are needed and, until this has been properly achieved, the Bank should be much more cautious in its prescriptions for the forest sector of these countries. The same concerns apply to boreal and dry tropical forests.

12. Poor analysis of Forest Conservation Projects

The 1991 Forest Policy places a welcome emphasis on Forest Conservation. However the review ignores the balance of payments problems incurred by borrower countries taking out loans to finance national parks management.

The returns, in foreign currencies, from these investments may be very long in coming. On the other hand, while the review mentions many GEF biodiversity projects-which are grant funded-it fails to even mention the critical conclusions drawn by the GEF Review of the severe problems with the GEF's biodiversity projects. In particular, the review does not examine the serious problems resulting from GEF funding being directed by the flawed concept of incremental costs for global benefits.

Moreover, whereas the Forest Policy prioritises novel conservation planning to involve local communities, the review does not evaluate progress towards achieving this goal and avoids mentioning the forced relocation of forest dwellers in recent Bank and GEF conservation projects.

13. Flawed Approach to Plantations

The review does summarise the efforts made since 1991 to build up plantations in tropical forest countries (under the rubric of 'Resource Expansion').

However, the review avoids confronting some of the key questions raised about plantations in recent years such as: the extent to which plantation establishment contributes to the conservation of natural forest (or not); whether plantations pose problems to local customary rights holders or contribute to local peoples' welfare and whether or how local, subsistence and outside commercial interests can be reconciled; what are the criteria for conversion of either natural or degraded forests, or other lands to plantations. Given the importance that plantation establishment is likely to play in future Bank lending (especially IFC investment) a much more detailed treatment of these issues is necessary.

14. Unsubstantiated Correlations with Forest Valuation

The review also perpetuates some highly questionable assumptions about forest economics which continue to adversely skew Bank lending. In particular, in line with earlier Bank thinking, the review makes the assumption that increased valuation of forests, reflected in higher prices for forest products and higher stumpage etc., will encourage more prudent forest management.

No evidence is presented to substantiate this assumption and the review ignores notable national examples where increased forest valuation has accelerated unsustainable logging and consequent forest clearance.

In the same way, the review maintains a doctrinaire position on the undesirability of log export bans, based both on a lack of examination of the social and environmental costs and benefits of such bans and a lack of appreciation of the way increased border prices relate back to forest extraction and the build up of domestic processing capacity. Far from improving logging and processing efficiency, the premature lifting of log export bans risks the collapse of domestic processing industries and the opening up of poorly regulated forests to uncontrolled extraction by foreign companies. The review fails to address these questions.

15. No Explanation of Policy and Institutional Reform Failures

Although the Bank reiterated its belief that policy reform, rather than investment, should be its priority in supporting improved forest management, the draft review notes that it has so far failed to implement any approaches for identifying or promoting reforms. It further acknowledges that its support for institutional reforms has been meagre, and has sometimes supported counter-productive institutional activities. Nevertheless, the review does not examine why these failures have occurred and what remedies it proposes to rectify the situation.

16. Inadequate Attention to Non-timber Forest Products

As long as the Bank continues to refer to NTFPs as 'minor' forest products, it will continue to give them insufficient attention in project development, as it has in the review. This suggests that the Bank has yet to develop a holistic approach to forests. The review needs to adopt a multi-purpose perspective, recognising the manifold values of forests.

Recommendations

We would like to stress that we welcome the evident willingness of the Bank to engage in open dialogue about the review through its consultations with NGOs. We were, in particular, impressed by the frank and constructive attitude of the Central Forestry Team, which made clear at the close of the consultations that it is seeking further input from NGOs to deepen and broaden its analysis and improve the review.

We look forward to contributing further to this process and hope you will construe this letter as a genuine attempt to improve the review and, thereby, the Bank's forest-related lending.

At the same time it is clear that the shortcomings of the review as it stands are not minor and it will require substantial further work if they are to be adequately addressed so that the review can usefully contribute to the evolution of Bank policy towards forests.

We thus urge:

1. That the date for the submission of the review to the Executive Directors be delayed to allow time for an adequate revision.
2. That the issues highlighted above, as well as others also raised during the consultations, are fully dealt with in the revision process.
3. That the Review expand its scope to include an assessment of the degree to which non-forest sector lending, including loans for macro-economic reform and structural adjustment, has adhered to the policy.
4. That the review team carry out at least a sample of field evaluations, in collaboration with local NGOs and peoples organisations, to gauge the accuracy of the available information.

5. That in the meantime the Board makes clear to Bank operational staff the pressing need for them to adhere scrupulously to the existing policy in both forestry and non-forest sector lending. Bank staff need to be forcefully reminded of the need to adopt a precautionary approach in all forest related lending.
6. That the Bank extends its prohibition on the financing of logging in old growth forests to dry tropical, temperate and boreal forests, which are equally vulnerable to abusive practice.
7. That the Bank establishes a permanent and independent mechanism to monitor and evaluate its forest-related activities.

2

Forest Legacy Program Implementation Guidelines

INTRODUCTION

The Cooperative Forestry Assistance Act (CFAA) of 1978, as amended, (16 U.S.C. 2101 et. seq.) recognizes that the majority of the Nation's productive forest lands are in private ownership; that private landowners are facing increased pressure to convert their forest lands to other uses; that greater population density, user demands and restrictions on Federal and other public lands are placing increased pressures on private lands to provide a wide variety of products and services from working forests including timber and other forest commodities, fish and wildlife habitat, watershed function and water supply, aesthetic qualities, historical and cultural resources, and recreational opportunities; and that good stewardship of privately held forest lands requires a long-term commitment that can be fostered through a partnership of Federal, State, local government and individual efforts.

In 1990, the Forest Legacy Program (FLP) was established to promote the long-term integrity of forestlands. The Secretary was directed to establish the FLP in cooperation with State, regional, and other units of government.

In carrying out this mandate, the Secretary of Agriculture is authorized to acquire lands and interests in lands in perpetuity for inclusion in the FLP. Landowner participation in the FLP, including the sale of lands and interests in lands, is entirely voluntary. The Program is implemented through State participation, consistent with these National FLP guidelines, and as described in each State Assessment of Need.

The FLP goals and objectives are accomplished through Forest Service (FS) cooperation with State partners, Federal agencies, local units of government, forest landowners and other partners. The FLP identifies and protects environmentally important private forestlands that are threatened by conversion to nonforest uses and provides the opportunity for continuation of traditional forest uses, such forest management activities and outdoor recreation.

The guidelines are organized in three parts:

PART 1-General Program Guidelines: Program direction applicable to all aspects of the FLP.

PART 2-State Grant Program Guidelines: Program direction applicable to States and Forest Service (FS) Regions/Area/IITF where a State has elected the State grant option and where ownership of lands or interests in lands is vested in a State or subdivision of a State.

PART 3-Federal Acquisition Program Guidelines: Program direction applicable to States and FS Regions/Area/IITF selecting the Federal acquisition and ownership process, where ownership of lands or interests in lands is vested in the United States (U.S.).

GENERAL PROGRAM GUIDELINES

Authority and Purpose of the Forest Legacy Program (FLP)

A. Authority

The Cooperative Forestry Assistance Act (CFAA) of 1978, as amended, (16 U.S.C. 2101 et. seq.) provides authority for the U.S. Secretary of Agriculture (Secretary) to provide financial, technical, educational, and related assistance to States, communities, and private forest landowners. Section 1217 of Title XII of the Food, Agriculture, Conservation and Trade Act of 1990 (P.L. 101-624:104 stat.3359; 16 U.S.C. 2103c), also referred to as the 1990 Farm Bill, amended the CFAA and directs the Secretary to establish the FLP to protect environmentally important forest areas that are threatened by conversion to nonforest uses. This authority continues indefinitely. Through the 1996 Farm Bill (Federal Agricultural Improvement and Reform Act of 1996; P.L. 104-127; Title III-Conservation; Subtitle G-Forestry; Section 374, Optional State Grants for Forest Legacy Program), the Secretary is authorized, at the request of a participating State, to make a grant to the State to carry out the FLP in that State, including the acquisition by the State of lands and interests in lands.

B. Purpose of the Forest Legacy Program

The purpose of the FLP is to ascertain and protect environmentally important forest areas that are threatened by conversion to nonforest uses.

FLP seeks to promote forestland protection and other conservation opportunities. Such purposes shall include the protection of important scenic, cultural, fish, wildlife and recreational resources, riparian areas and other ecological values. Traditional forest uses, including timber management, as well as hunting, fishing, hiking, and similar recreational uses are consistent with purposes of the FLP. Both purchased and donated lands and interests in lands through the use of conservation easements and fee-simple purchase are used to acquire forested land meeting Forest Legacy purposes from willing sellers or donors.

C. Delegations of Authority

The Secretary has delegated authority to administer all aspects of the FLP to the Under Secretary for Natural Resources and Environment (7 CFR 2.20(a)(2)(xvi)) who in turn has delegated the authority to the Chief of the Forest Service (7 CFR 2.60(a)(16). Delegations only apply within the U.S. Department of Agriculture and its agencies. The role of State and Regional programs, and the right of States to elect the State Grant Option, are contained in the authorizing statute and these program implementation guidelines.

Description of Terms and Abbreviations

Assessment of Need (AON) is a document produced by a State, or a federally recognized Indian Tribe, in consultation with the State Forest Stewardship Coordinating Committee (SFSCC). The AON contains the an assessment of the forests and forest uses, a description of forces that are converting forests to nonforest uses, describes Eligibility Criteria developed by the State to identify important forest areas to be proposed as Forest Legacy Areas (FLA), and acts as a guide to implementation of FLP in the State.

Assessment of Need (AON) Amendment is a document produced by a State to amend their AON, to add or delete Forest Legacy Areas (FLA), or to modify the Eligibility Criteria. CFAA is the Cooperative Forestry Assistance Act of 1978, P.L. 95-313, 92 stat. 365, 16 U.S.C., 2101 et seq. (as amended through P.L. 107-195, June 16, 2002).

Conservation Easement is a legal agreement a property owner makes with a governmental entity or a nonprofit organization to restrict activities allowed on the land in order to protect specified conservation values. Easement restrictions are tailored to the particular property and to the interests of the individual landowner. All FLP conservation easements are held in perpetuity.

Eligibility Criteria are a set of factors developed by the State lead agency, in consultation with the State Forest Stewardship Coordinating Committee (SFSCC), to evaluate geographic areas to determine if they contain significant environmental values to be considered an 'important forest area' and contain "threats" of conversion to be eligible as a Forest Legacy Area (FLA).

Federal Appraisal Standards are those standards contained in the publication entitled "Uniform Appraisal Standards for Federal Land Acquisitions: Interagency Land Acquisition Conference, 2000," or subsequent amendments or updates.

Forest Legacy Area (FLA) is a geographic area with important forest and environmental values, that satisfies identified Eligibility Criteria and has been delineated, described, and mapped in a State's AON for the FLP. Acquisition of lands and interests in lands for the FLP can only occur within approved FLAs.

Forest Legacy Area (FLA) Boundary Adjustment is a minor change to an existing FLA to create a more logical or manageable boundary.

Forest Legacy Program (FLP) Project is an individual or series of land or interest in land acquisition transaction(s). The transaction(s) can be on an individual tract or multiple tracts in a distinct geographical area. A FLP project relates to a single funding event in a given fiscal year. FLP projects can have a single parcel that can be completed at one closing or more than one parcel that can be completed in a succession of closings. If a successive FLP project is proposed on a parcel or in a distinct geographic area each transaction is treated as an independent unit in the project selection process and funding is not guaranteed.

Forest Service (FS) is the United States Department of Agriculture Forest Service. Forest Service Region/Area/IITF refers to the field units of the Forest Service responsible for FLP management and oversight within the Forest Service Regions, Northeastern Area (Area) or International Institute of Tropical Forestry (IITF).

Forest Stewardship Plans, or multi-resource management plans, are prepared with the purpose of achieving long-term stewardship of forestland. Such plans identify landowner objectives and describe actions to protect and manage soil, water, range, aesthetic quality, recreation, timber, and fish and wildlife resources, and other conservation values identified on the tract. Plans are to be prepared by a professional resource manager. A Forest Stewardship Plan that meets the requirements of the Forest Stewardship Program or a multi-resource management plan is required for FLP qualification. The State Forester or equivalent, or their designee must approve the plan.

Full Fee Purchase is a land conveyance where a purchaser acquires all rights, title and interest in a property from a seller or owner. It is also known as fee simple or fee acquisition. Geographic Regions are the collection of States that makeup the National Association of State Foresters (NASF) Regions. The three regions are: North (consisting of the States within the FS Northeastern Area), South (consisting of all the States within the FS Southern Region, and the Territories of the International Institute of Tropical

Forestry), and the West (consisting of all the States within the FS Northern, Rocky Mountain, Intermountain, Southwestern, Pacific Southwest (including the Commonwealth of the Northern Mariana Islands, Guam and American Samoa), Pacific Northwest and Alaska Regions.

Indirect costs relate to costs of the management and administration of the FLP. Indirect costs, unlike salary, which is a direct cost, are defined as costs not readily assignable to a specific legacy acquisition.

In-kind contributions are non-cash contributions, including third-party contributions. In-kind contributions must be expenses necessary to accomplish program activities, and allowable if the Federal Government were required to pay for them.

Interests in Land are a right, claim, or legal share in real property that are less than the full title. Land Trust is a nonprofit organization, as described in 501(c) of the Internal Revenue Code of 1986, that protects land by working with landowners who wish to donate or sell fee title or conservation easements to maintain conservation values associated with the land.

Market Value is the amount in cash, or in terms reasonably equivalent to cash, for which in all probability the property would be sold by a knowledgeable owner willing but not obligated to sell to a knowledgeable purchaser who desires but is not obligated to buy. (Uniform Appraisal Standards for Federal Land Acquisitions: Interagency Land Acquisition Conference, 2000, p.4.) Multi-State Entity is a government-established organization involving two or more States or Indian tribes whose jurisdiction encompasses all or portions of the land area of an FLA(s). National Association of State Foresters (NASF) is the organization representing State forestry organizations in all 50 States, the territories and the District of Columbia.

Nonfederal Cost Share refers the nonfederal cost-share required to receive FLP funding. There are three main categories of activities that meet this requirement:

(1) the value of land, or interests in land, dedicated to the FLP that is not paid for by the Federal government,

(2) nonfederal costs associated with program implementation, and

(3) other nonfederal costs associated with a grant or other agreement which meets FLP purposes. The nonfederal cost-share must be documented, and in the case of a grant, must meet the timing, terms and conditions of the grant.

Nonforest Uses

Noncompatible-nonforest uses are uses of the land inconsistent with maintaining forest cover including, but not limited to, activities that result in extensive surface disturbance such as residential subdivisions, commercial development, and mining. These uses generally should be excluded from FLP conservation easements or land purchases. FLP funds should only be used on parcels with forestland as defined in the State's AON. Compatible-nonforest uses are nonforest uses of the land that may be compatible with forest uses as part of an undeveloped landscape, including cultivated farmland, pasture, grassland, shrubland, open water, and wetlands. These nonforest uses should be less than 25 percent of the total area. Forest Legacy funds should only be used on parcels with forestland as defined in a State's AON. Other funding sources may be used to protect nonforested areas on those parcels with less than the minimum required forest cover.

Nontrust Allotment Lands are privately owned fee simple lands owned by tribal members and if they are forested, are eligible for the FLP when they are located within an approved FLA. Trust lands and reservations are already protected through the trust relationship between the U.S. Department of the Interior and the tribe and are ineligible for the FLP.

Pass-through describes a land transaction whereby a third party, such as a land trust, acquires interests in lands with the intent to convey such interests to a unit of government. The transaction can include a full or partial donation, or sale at market value.

Payment in Lieu of Taxes (PILT) is made by tax-exempt entities, including the Federal government, to compensate local jurisdictions for tax revenues foregone as a result of ownership by a tax-exempt owner. Any FLP tract acquired in fee and held by the FS is eligible for PILT payments (entitlement land as defined at 31 U.S.C. 6901). Federal funds for PILT are not authorized for any land or interests in land held by nonfederal entities, or for conservation easements held by the United States.

Program Funds are FLP funds that are appropriated by Congress and allocated by the FS to three categories: Project funds, Administrations funds, and AON Preparation funds. Project Evaluation Criteria are developed by the States, in consultation with the State Forest Stewardship Coordinating Committees (SFSCC), to evaluate the eligible tracts submitted by interested landowners for inclusion in the FLP.

Relocation refers to the provision in the Uniform Relocation Assistance and Real Estate Property Acquisition Policies Act of 1970 (PL 91-646 or 42 U.S.C. 4601) which requires Federal agencies and programs to pay for the relocation of a person displaced by a federally funded real estate transaction.

Reserved Areas are designated areas where nonforest uses (e.g. house, barn, remote recreation camps, etc.) are or will be allowed, but are inseparable from the land holding and do not have a detrimental effect on the conservation easement values. These areas shall be defined and described in the conservation easement and may be restricted in terms of their use, or provisions made through cost and time to cure and treatment. To the extent possible these areas of non-compliance should be excluded from the FLP project.

Reserved Interest Deed is where the grantee (government) acquires all rights, titles, and interests in a property, except those rights, titles, and interests that may run with the land that are expressly reserved by a grantor (landowner).

Secretary is the U.S. Secretary of Agriculture.

State refers to any of the 50 States, Puerto Rico, Guam, the United States Virgin Islands, the Commonwealth of the Mariana Islands, and American Samoa participating in the FLP. State Forest Stewardship Coordinating Committees (SFSCC) are defined, and their duties are described, in Section 19(b) of the CFAA (16 U.S.C. 2113). They are chaired and administered by the State Foresters, or equivalent State officials, with membership composed of representatives from the following agencies, organizations, or individuals: Forest Service; Natural Resources Conservation Service; Farm Services Agency; Cooperative State Research, Education, and Extension Service; local government; consulting foresters; environmental organizations; forest products industry; forest land owners; land trusts; conservation organizations; the State fish and wildlife agency; and others determined appropriate by the Secretary.

The SFSCC makes recommendations to the State lead agency regarding the AON, AON amendments, and the determination of project priorities.

State Lead Agency is that unit of State government responsible for coordinating the establishment and implementation of the FLP in the State, as designated by the Governor or pursuant to State law. The State lead agency is usually a forestry agency, but may be another natural resource agency.

Tribal Assessments of Need-An AON is developed by a federally recognized Indian Tribe in cooperation with the State and the SFSCC. Only nontrust allotment lands are eligible for FLP. Lands or interests in lands purchased under a Tribal FLP can be through a grant to a cooperating State or through the Federal acquisition option.

Working Lands Conservation Committee is a committee of the NASF having coordination and consultation responsibilities within that organization regarding the FLP.

National Environmental Policy Act (NEPA)

NEPA applies to certain proposed actions of the Federal Government. NEPA does not apply to the independent actions of States or private property owners. It has no applicability to a private property owner's use or development of his/her property rights, nor the development of a State's FLP. It could apply to Federal agency actions undertaken on private property if the U.S.

acquired a right to permit or deny certain land uses and then proposed to exercise that right, but in such an instance it would be the U.S. that would be required to satisfy NEPA requirements, not the private owner. It should be known that:

1. A Programmatic Environmental Assessment and a Finding of No Significant Impact was completed for the national FLP and signed by the Chief of the Forest Service.
2. Under the Federal acquisition option, the FS NEPA regulations (Forest Service's Environmental Policy and Procedures Handbook 1909.15-92.1, effective 9/21/92), the acquisition of an individual Forest Legacy tract and/or easement may be categorically excluded from the preparation of an Environmental Impact Statement or an Environmental Assessment unless scoping indicates extraordinary circumstances exist.

Coordination with State Forestry Agencies

Whereas most State lead agencies are State Forestry agencies, and the CFAA establishes a broad cooperative relationship between the FS and State Foresters, the FS shall appoint a representative to coordinate with the Working Lands Conservation Committee of the NASF (or its successor) regarding the FLP. Periodically, the Director of Cooperative Forestry, and the appointed FS representative shall meet with the NASF Working Lands Conservation Committee to assess program operations, accomplishments, and policies. In States where the State Forestry agency is not the designated State lead agency for the FLP, a coordinating mechanism shall be instituted between the State lead agency, the State Forester, and the SFSCC.

Assessment of Need (AON) and Identification of Forest Legacy Areas (FLA)

A State or a federally recognized Indian tribe conducts an AON, in cooperation with the SFSCC, to document their need for inclusion in the FLP, through an evaluation of current forests, forest uses, and the trends and forces causing conversion to nonforest uses. Federally recognized Indian Tribes must cooperate with the SFSCC when conducting an AON for nontrust allotments lands. The AON is intended to define the Eligibility Criteria to be used in the identification of important forest areas to be proposed as an FLA; identify and delineate the boundaries of forest areas meeting the Eligibility Criteria for designation as an FLAs; determine through analysis what defines "threatened" and "environmentally important forests"; and outline the State's project evaluation and prioritization procedures. The AON must be developed in consultation with SFSCC and approved by the State lead agency.

State lead agencies may utilize the services of land trusts or other entities in preparing the assessment. Information from existing sources may be used to prepare the AON, instead of initiating new studies that would duplicate existing data. Examples of appropriate sources include State Forest Resources Plans, State Comprehensive Outdoor Recreation Plans, growth management studies, State cultural site inventories, inventories of threatened and endangered species, and other State, regional or local plans, studies or reports. The AON shall include relevant information about both public and private lands, address the issue of how best to maintain the integrity of forestlands for future generations, and address pertinent issues as identified by the State.

At a minimum, the AON must address the following as they relate to the purpose of the FLP:

1. Forest resources including:
   - o Aesthetic and scenic values;
   - o Fish and wildlife habitat;
   - o Minerals resource potential;
   - o Public recreation opportunities;
   - o Soil productivity;
   - o Forest products and timber management opportunities;
   - o Watershed values including water quality protection;
2. The present and future threat of conversion of forest areas to nonforest uses. States are responsible for defining the conversion threat(s);

3. Historic uses of forest areas, and trends and projected future uses of forest resources;
4. Current ownership patterns and size of tracts, and trends and projected future ownership patterns;
5. Cultural resources that can be effectively protected;
6. Outstanding geological features;
7. Threatened and endangered species;
8. Other ecological values;
9. Public recreational opportunities;
10. Protected land in the State, to the extent practical, including Federal, State, and municipal lands and land trust organizations lands;
11. Issues identified by the SFSCC and in the public involvement process.

Using the above information the AON shall include the following:

1. Identification of applicable Eligibility Criteria;
2. Identification of specific FLA(s) for designation;
3. Specific goals and objectives to be accomplished by the FLP;
4. Process to be used by the State lead agency to evaluate and prioritize projects to be considered for inclusion in the FLP.

The project evaluation and prioritization process outlined in the AON should reflect the direction set forth in the CFAA to give priority to lands which can be effectively protected and managed, and which have important scenic or recreational values, riparian areas, fish and wildlife values including threatened and endangered species, or other ecological values. Traditional forest uses such as forest management activities, including timber management, and outdoor recreation opportunities are considered consistent with purposes of the FLP and are encouraged on FLP tracts when consistent with the State's AON and the conservation purposes for FLP tract acquisition. The prioritization process should implement a strategy that enhances existing protected forestlands or local and State conservation strategies as outlined in the AON.

The composition of the SFSCC is defined in Section 19(b) of the CFAA (16 U.S.C. 2113). States are encouraged to broaden this composition to include interests appropriate to benefit the FLP. This committee cooperates with the State lead agency in the preparation of the AON, identification of FLA Eligibility Criteria, the identification of proposed FLAs from which lands may be entered into the FLP, and recommendation of priority lands to be considered for enrollment in the Program.

Public participation and involvement in the AON preparation is a State responsibility. In the absence of established State procedures, NEPA may serve as an appropriate model for public involvement. The State lead agency will solicit involvement and comments on the AON from the public including State and local governments. The goals of public involvement include hearing concerns and views from interested and affected individuals and organizations, receiving new information, identifying and clarifying issues.

Based on the State-wide AON, the State lead agency, in consultation with the SFSCC, identifies specific geographic FLAs that meet the Eligibility Criteria, and recommends them to the FS for designation as of a FLA.

States are encouraged to cooperate in the identification of FLAs that cross State boundaries and to work together to coordinate acquisitions of lands or interests in lands that have complementary purposes. However, program implementation is undertaken by the individual States (State Grant Option) or by the FS (Federal Option).

The identification of proposed FLAs must include:

1. Location of each geographic area on a map and a written description of the proposed FLA boundary;
2. Summary of the analysis used to identify the FLA and its consistency with the Eligibility Criteria;
3. Identification of important environmental values, and how they will be protected and conserved;

4. The conservation goals or objectives in each FLA
5. List of public benefits that will be derived from establishing each FLA;
6. Identification of the governmental entity or entities that may hold lands or interests in lands (State grant option) or may be assigned management responsibilities for the lands and interests in lands enrolled in the program (Federal option); and
7. Documentation of the public involvement process and analysis of the issues raised.

Eligibility Criteria for Establishing Forest Legacy Areas (FLAs)

The CFAA directs the Secretary to establish Eligibility Criteria for the designation of FLAs, in consultation with the SFSCC. These criteria should be based upon the FLP purpose to protect environmentally important forest areas that are threatened by conversion to nonforest and be further developed through the AON.

FLA boundaries must encompass forestlands with significant environmental and other resource-based values. Areas may also include nonforested areas such as farms and villages if they are an integral part of the landscape and are within logical boundaries. Since FLA boundaries may not correspond to property boundaries, tracts located partially within the geographically defined FLA are eligible for the FLP, upon approval of a boundary adjustment by the FS Region/Area/IITF.

Indian reservations and tribal lands may have important features on the forested landscape. Indian tribes and States are encouraged to collaborate and to consider only nontrust allotment lands for designation as, or inclusion within, a FLA. Other tribal lands are already protected through the trust relationship between the U.S. Department of the Interior and the tribe and are ineligible for the FLP.

States are responsible for determining what defines "threatened" and "environmentally important forest areas" in the State. However, environmentally important forest areas shall contain one or more of the following important public values, as defined by the States:

1. Timber and other forest commodities
2. Scenic resources;
3. Public recreation opportunities;
4. Riparian areas;
5. Fish and wildlife habitat;
6. Known threatened and endangered species;
7. Known cultural resources;
8. Other ecological values.

The FS, State or unit of State or local government may only acquire lands and interests in lands identified within a FLA under FLP authority on a willing seller/willing buyer basis.

AON and Amendment Approval

The State lead agency must submit the AON, including proposed FLAs and Eligibility Criteria, to the FS Region/Area/IITF. The FS Region/Area/IITF with input from the FS Washington Office reviews the AON and works with the State lead agency to complete the AON. Once finalized, the FS Washington Office forwards the AON to the Secretary for final approval. Final approval establishes the FLP for the State.

AONs shall be periodically reviewed (at least at 5-year intervals) by the State lead agency, the FS Region/Area/IITF, and the SFSCC to assess whether AON amendments or updates are necessary. The results of reviews will be documented by the State lead agency. AONs should be amended as needed.

The State lead agency may amend the AON to make significant changes or minor adjustments. Significant changes include modifications to their FLP, changes to the FLA Eligibility Criteria, or to add or delete a FLA.

These changes need to be made in consultation with the SFSCC and with public involvement. FLAs and project evaluation criteria shall be of a scale and detail to effectively focus delivery of the FLP.

Significant Amendments to an AON may address the following:

1. Issues associated with maintaining the integrity of forestland and the proposed FLA specifically.
2. Revision, if any, of the FLA Eligibility Criteria.
3. Changes in policies or conditions that have occurred since the previous AON;
4. The identification of proposed FLA(s) and conservation goals or objectives associated with that FLA.

The Chief of the Forest Service, or designee, provides final approval of the Amendment of an AON authored by a State lead agency after consultation with SFSCC. In addition, the State lead agency may complete minor AON amendments, such as FLA boundary adjustment or project prioritization process. These minor changes need to be coordinated with the SFSCC, and need review and approval by the appropriate FS Region/Area/IITF.

Multi-State Identification of Forest Legacy Areas

States are encouraged to cooperate in the identification of FLA that cross State or Tribal boundaries and to work together to coordinate acquisitions of lands or interests in lands that have complementary purposes. States may elect to jointly use an existing or new multi-State or regional entity to identify FLAs or develop FLP projects that cross State boundaries. The entity must be a government-established organization, whose jurisdiction encompasses all or portions of the land area of the FLA States involved. However, program implementation is undertaken by the individual States (State Grant Option) or by the FS (Federal Option). The entity conducting a multi-State identification of FLAs is responsible for:

- o Obtaining approval from the appropriate States or Indian tribes for FLAs within their boundaries,
- o Cooperating with appropriate SFSCCs,
- o Obtaining public comments on the identification of FLAs, and
- o Complying with all other requirements of these guidelines.

Project Selection Process

The FS will conduct a project selection process to arrive at a prioritized national project list for consideration in the President's budget for the upcoming fiscal year. The project selection process and calendar of due date milestones are developed in consultation with the State lead agencies and FS Region/Area/IITF and communicated by the FS Washington Office. The FS will ensure that national evaluation and prioritization criteria are communicated to the States and in a timely manner so that submitted projects adhere to strategic goals and objectives of FLP. Project selection steps are:

Step 1: Release Project Selection Calendar with Due Dates

The project selection process and calendar of due date milestones are developed in consultation with the States and FS Regions/Area/IITF and communicated by the FS Washington Office.

Step 2: State Project Prioritization and Submission

FLP project applications are accepted by the State lead agency as outlined in the State's AON. The SFSCC reviews and evaluates applications according to the criteria identified in the State's AON, authorizing statute, and other relevant direction and policy, and provides recommendations to the State lead agency. Projects approved and prioritized by the State lead agency are forwarded to the FS Region/Area/IITF for funding consideration. Only projects submitted through this process will be deemed eligible. Each State will prepare a list of projects and enter or update its list for submission to the FS via the Forest Legacy Information System or other means as requested.

Step 3: Forest Service Regional Review

FS Regions/Area/IITF will review submitted projects considering State priorities and national criteria. The purpose of this review is to improve project viability, facilitate the national project selection process and advance the strategic outcomes of the FLP. FS Regions/Area/IITF will submit projects to the FS Washington Office for funding consideration.

Step 4: National Review; Develop National Project List

The FS Washington Office will develop a prioritized national project list by convening a panel. There are 3 purposes of the panel;

(1) assure that all projects meet Congressional and Administration direction;

(2) assure that projects meet national program goals; and

(3) develop a National List of ranked projects.

The composition of the panel shall be developed annually in consultation between the State lead agencies and the FS, and will be representative of geographic regions. Project evaluation and ranking is based on the following national core criteria; project readiness will be considered as well as other evaluation considerations developed in consultation with State lead agencies and FS Regions/Area/IITF. The national core criteria are:

- o *Important:* The public benefits gained from the protection and management of the property including environmental values, and the economic and social aspects;
- o *Threatened:* Conversion to nonforest uses is likely or imminent and will result in a loss of forest values and public benefits; and
- o *Strategic:* Fits with a larger conservation plan, strategy, and initiative and enhances previous conservation investments.

States newly entering FLP will be given a "New-State start-up" preference for an initial FLP project. This is a placeholder for planning purposes and does not guarantee project funding. In order to receive the New State start-up project funds the State must have an approved AON and the project must meet national core criteria and the State's evaluation criteria and be submitted within the fiscal year that the placeholder is approved by Congress.

Step 5: Submit National FLP Project List to the Office of Management and Budget and to Congress

Each fiscal year, the FS Washington Office will submit a project list to the Office of Management and Budget for funding consideration in the President's budget. Once the President's budget has been completed, the FS Washington Office will notify the appropriate House and Senate Committees and Subcommittees of the recommended projects for the upcoming Fiscal Year.

PROGRAM FUND CATEGORIES

Forest Legacy funds are allocated to one of three categories: Project Funds, Administration Funds, and AON Preparation Funds. FLP funds may not be used for monitoring and enforcement.

Project Funds

Project funds are those used to directly purchase lands or interests in land joining the FLP. Project funds may be expended by the State lead agency or the FS, as applicable, to cover transaction costs, including but not limited to: appraisals and appraisal review, land surveys, closing costs, establishing baseline information, title work, purchase of title insurance, conservation easement drafting, and other real estate transaction expenses for those tracts. Project funds may also be expended to facilitate donations of land or interests in lands to a qualified and willing donee for FLP purposes, by paying for expenses directly related to the donation, including but not limited to, land surveys, conservation easement drafting, title work, and establishing baseline information. For an outright donation of a conservation easement or land, FLP program funds may not be used to pay for an appraisal. In the case of a partial donation of a conservation easement or land, an appraisal meeting Federal standards is required to determine the value of property. FLP funds may be used for appraisals on these partial donations. When Federal funds are used to purchase real property, including conservation easements, appraisal and acquisition work procedures must meet Federal standards.

Administration Funds

Administrative funds are the portion of funds used for day-to-day program management at all levels. Administration funds may be used for a variety of activities, including FLP program administration, personnel and overhead, and all activities identified as eligible uses of project funds to prepare projects and potential projects. Forest Legacy funds for administration shall be kept to a minimum. As a goal, all attempts should be made to keep administration funds under 15 percent of the total funds appropriated.

AON Preparation Funds

AON preparation funds may be made available to States to help defray the cost of preparing, or amending an AON.

Process for Allocating Funds to Forest Service Regions/Area/IITF

Following passage of the annual appropriations bill, the FS Washington Office develops the Forest Legacy Program Direction and allocates funds to the FS Regions/Area/IITF for distribution. The allocation process differs for each fund category described below.

Allocation of Project Funds

Allocations to FS Regions/Area/IITF are based on the results of the national project selection process and the appropriations bill. Under the State grant option, FS Regions/Area/IITF will award grants to States for specific, identified projects.

Allocation of Administration Funds

The FS Washington Office distributes administration funds to FS Regions/Area/IITF. Each FS Region/Area/IITF in consultation with the States requests these funds to meet their needs and the needs of the participating states in their Region/Area/IITF. Administration funds are also used by the FS Washington Office to fund program management functions. Administration funds will be granted to States under the State grant option separately from project funds.

Allocation of AON Preparation Funds

The FS Washington Office distributes AON preparation funds to the States by way of FS Regions/Area/IITF. These funds are requested by FS Regions/Area/IITF to meet the needs of their States to develop new AONs or amendments.

Redirection and Reprogramming of Funds

Due to the nature of real estate transactions, FLP projects may change in scope, cost or fail completely. These changes can result in unspent or excess funds for some projects while others may need additional funding to bring them to completion.

In order to maximize the efficient and effective use of FLP project funding, the FS will either redirect or request reprogramming of funds. Redirection is a shift of funds from one congressionally approved project to one or more other congressionally approved project(s). Reprogramming is a shift of funds that exceeds an increase or decrease of 10% per project not to exceed $500,000 to an existing project, or shifting of any amount of funds to a project not previously approved by Congress.

Regional Redirection Process

FS Regions/Area/IITF may redirect up to an increase or decrease of 10% per project not to exceed $200,000 of project funds that are excess or unspent from one project to one or more other Congressionally approved project(s) within the FS Region/Area/IITF which is underfunded and where there is a substantiated need (e.g. loss of other funding sources, appraisal documenting increased cost, etc.) to bring the project to completion.

Project funds over $200,000, or those that cannot be redirected by the FS Region/Area/IITF, will be released for the national process. FS Regions/Area/IITF will notify the FS Washington Office before a redirection takes place and report these actions periodically. All funds from failed projects will be released for the national process.

National Redirection Process

The FS Washington Office, through consultation with FS Regions/Area/IITF, may redirect up to an increase or decrease of 10% per project not to exceed $500,000 of project funds that are excess or unspent from one project to one or more congressionally approved project(s) which is underfunded and where there is a substantiated need (e.g. loss of other funding sources, appraisal documenting increased cost, etc.) to bring the project to completion. In addition, when funds have not been spent or contractually obligated within two years of receipt of funds, they revert to the FS Washington Office via the appropriate FS Region/Area/IITF. The FS Washington Office will:

1. Assess the extent of unspent or returned funds on a periodic basis;
2. Facilitate selection and funding of underfunded projects not addressed by the regional process or between FS Regions/Area/IITF; and
3. Notify Appropriations Subcommittees of any redirection action taken by the FS FS Regions/Area/IITF or FS Washington Office.

National Reprogramming Process

The FS Washington Office, through consultation with FS Regions/Area/IITF, may request reprogramming by the Appropriations Subcommittees of unspent or returned funds to a project that requires more than an increase or decrease of 10% per project not to exceed $500,000 to complete. In addition, the FS Washington Office may request reprogramming by the Appropriations Subcommittees of unspent or returned funds to a project not previously approved by Congress. The FS Washington Office will:

1. Determine the funds available for reprogramming on a periodic basis.
2. Identify underfunded projects that cannot be addressed through the Regional Redirection Process and determine the priority for reprogramming.
3. Recommend reprogramming to fund projects from the National Project List next in sequence in priority ranking to the extent practicable.
4. Submit reprogramming requests to the Appropriations Subcommittees for approval.
5. Allocate funds to projects approved for reprogramming.

FLP Cost Share Requirements

The CFAA directs that, to the extent practicable, the maximum Federal contribution for total program costs may not exceed 75 percent. To assure program-wide cost share goals are met, each project budget must include a minimum nonfederal contribution of 25 percent. This nonfederal cost-share must meet Forest Legacy purposes. It may consist of:

(1) the value of land, or interest in land, dedicated to the FLP that is not paid for by the Federal government;

(2) nonfederal costs associated with program implementation; and

(3) other nonfederal costs associated with a grant or other agreement that meets FLP purpose.

The nonfederal cost-share must be documented, and in the case of a grant, must meet the timing, terms, and conditions of the grant. The cost-share can occur at any phase of the FLP including planning, developing future projects, acquisition, capital improvement, management, or administrative activities. When a grant is involved, the cost-share must occur within the life of a grant and meet all grant requirements. Federal requirements identify the grant period as beginning when the grant is formally awarded and ends after two years to ensure that the federal funds are spent promptly. However, a grant may receive a maximum extension to five years. Allowable costs shall be determined in accordance with the 7 CFR 3016, "Uniform Administrative Requirements for Grants and Cooperative Agreements to State and Local Governments," and any amendments to this regulation.

Donations of land or interests in land must be documented to count as part of the nonfederal cost-share. The title does not need to be transferred to the State or federal government in order for the donation to qualify as cost share. However, if in the future, the donated lands are conveyed or the rights or title are modified in a way that is inconsistent with the purposes of the FLP then the State must restore the cost share value dedicated in the grant agreement. The value of donations may be included as part of the nonfederal cost-share if all of the following are met:

1. The donation contributes to the objectives and priorities of the State FLP as set forth in the AON;
2. All or part of the tract being donated must be within the boundaries of an FLA, and may include National Park, National Forest, National Wildlife Refuge, or other Federal land boundary, or within the boundaries of an area designated through an analogous State program with goals compatible with the FLP and be within an FLA;
3. The donor documents their desire that value of the interests may be used as cost share for the FLP project;
4. The donation of land or an interest in land must contain perpetual covenants to assure that the tract will be managed in a manner compatible with the goals for which the FLA was established;
5. The donee (holder of donated rights) is a unit of government or a non-profit conservation organization (land trust) that meets the eligibility requirements for holding a conservation easement established by the Internal Revenue Service and has as its purpose the management of lands or interests in land consistent with FLP purposes;
6. If the donation is in the form of a conservation easement then the deed needs to contain a provision that directs all of the easement holder's proceeds from a subsequent sale or exchange of interests in land be used in a manner consistent with the conservation purposes identified for the subject interests in lands;
7. The respective portion of the donation must not have been previously credited towards any Federal program's nonfederal cost share; and
8. The State lead agency approves the donation as contributing to the cost-share.

Acquisition of Lands or Interests in Lands

FLP acquisitions may be outright full fee purchases, or acquisition of development rights or other rights conveyed through a conservation easement. Except in the case of a full and complete donation of land or an interest in land, if any Federal funds are used in the acquisition of Forest Legacy tract the following shall apply:

1. Federal appraisal standards must be met, including appraisal review by a qualified Review Appraiser;
2. The landowner must be informed in writing of the market value and that sale of the property is strictly voluntary;
3. The landowner must be notified in writing that the property will NOT be purchased if negotiations do not result in amicable agreement;
4. Federal payment to the landowner for lands or interests in lands is not more than the market value as determined by an appraisal meeting Federal appraisal standards;
5. The title acquired must be free of encumbrances inconsistent with the purposes of the FLP. Title insurance may be secured for the full value of the encumbered property, but is not an alternative to an acceptable title; and
6. If relocation is involved the requirements in the Uniform Relocation Assistance and Real Estate Property Acquisition Policies Act of 1970 (PL 91-646 or 42 U.S.C. 4601) must be followed. The FS will be advised in advance of any acquisition involving relocation.
7. In the case of acquisition of interested in lands, the development of a Forest Stewardship Plan or multi-resource management plan that has been approved by the landowner and the State Forester or designee and Baseline Documentation Report shall be prepared prior to project closing (See Appendix J for sample content and references).

All FLP acquisitions of lands or interest in lands are perpetual and therefore run with the land. Although any remaining interests held by the landowner may be subsequently conveyed, future owners are still bound by the terms and conditions of the conservation easement. At the same time, future owners shall retain full control of the rights that are not acquired by the FLP, and shall be subject only to those restrictions that the present landowner has conveyed to the Federal, State, or local government. Compatible nonforest land uses (cultivated farmland, pasture, grassland, shrubland, open water, and wetlands) are desirable land uses in many FLAs. FLP funds should not be used for any property not meeting the State's definition of forested land in the AON, unless there is a written plan scheduling reforestation or afforestation. Programs to conserve farms, ranches and similar land uses may be used in conjunction with the FLP to protect properties where there are mixed forest and compatible nonforest uses.

Conservation easements are required to contain language pertinent to the purpose of the FLP and a reversionary provision to ensure the conservation investment of FLP into the future (Example clause language are found in Appendix I). During the development of tract specific conservation easements, a determination will be made as to whether the acquisition of mineral rights, prohibition on reserved areas, or an exclusion of the area that does not comply with FLP, would be necessary in order to protect the other rights that are being considered for acquisition. In some situations, it may be impossible to protect environmentally important forest areas pursuant to the purpose of the FLP without acquiring the mineral rights.

The FLP conservation easement holder (Federal, State or local government) is responsible to assure that baseline documentation contains all the information necessary to monitor, manage and enforce the easement. Where the conservation easement is a tax-deductible gift, and the owner retains rights to the property, the Internal Revenue Service (IRS) holds the donor responsible for providing sufficient baseline data "to establish the condition of the property at the time of the gift." (See Treas. Reg.§1.170A-14(g)(5)(i)). However, this does not eliminate the FLP need for baseline documentation.

Baseline documentation describes or depicts a tract of land and its attributes on the day it becomes restricted by an easement. This documentation is required on all FLP tracts and is completed prior to project closing. Documentation of the property should include a map of the area drawn to scale showing all existing man-made improvements or incursions such as roads, buildings, fences or gravel pits; an aerial photograph of the property taken as close to the date the property is restricted as possible; and on-site photographs, especially of significant features. The above should be accompanied by narrative descriptions of tract attributes and other pertinent information.

States and landowners are encouraged to display the official FLP signs on the FLP property using the signs in accordance with Appendix K. The posting of FLP tracts helps promote public awareness, recognition and support for the program. Landowner permission should be secured before posting any signs. Costs associated with sign posting can be covered by FLP project or administration grants or States may use such expenses as FLP cost share. Signs should be inspected during the annual monitoring of the FLP tract and repaired when in poor condition. FLP sign art and program logos may be used by FLP partners for items that contribute to the purpose of awareness (e.g. brochures, workshops, outreach efforts, posters, FLP information packets, web sites etc.)

Appraisal and Appraisal Review

The FLP policy on appraisal is that all FLP acquisition of land or interests in land using Federal funds must comply with Federal appraisal standards contained in the publication entitled "Uniform Appraisal Standards for Federal Land Acquisitions: Interagency Land Acquisition Conference, 2000," as amended or updated. Appraisals and appraisal reviews may be conducted by any qualified appraiser meeting the minimum standards outlined in Appendix H.

The FLP will ensure high quality appraisal service and accountability to the program by:

- o Annual planning and coordination of appraisal work to allow for efficient allocation of resources.
- o Requiring checks and balances:
- o States will ensure that qualified appraisers trained and competent in appraisal, appraisal review and

knowledgeable of Federal standards will be used. The State may use State, contract or Federal appraisal or review services to meet this requirement.

- o States or the FS will review contract appraiser qualifications as stated in Appendix H before they are employed to conduct a FLP project appraisal or review.
- o The appraiser and identified review appraiser will engage in an initial consultation before the project appraisal takes place. The review appraiser will develop project specific appraisal instructions for the appraiser as a result of this consultation.
- o The FS will conduct spot checks of appraisal reviews to ensure quality and accuracy.
- o Forest Legacy funds can only be used to purchase lands and interests in land after the appraisal review confirms that the appraisal meets the Uniform Appraisal Standards for Federal Land Acquisitions. It is recommended that an offer not be made until the appraisal review is approved.

Conservation Easement Monitoring, Management, Record-Keeping & Enforcement

The governmental entity holding title to interests in land acquired under the FLP shall monitor and manage those interests in perpetuity. The holder may delegate or assign monitoring, management, and enforcement responsibilities over lands and interests in lands acquired under the FLP only to other Federal agencies or State or local government entities.

Such delegation or assignment of responsibility shall be documented by a written agreement. The governmental entity responsible for monitoring, management and enforcement of the conservation easement may in turn delegate or assign management and monitoring authority to other parties, to include land trusts, conservation groups, and other governmental entities. Such delegation or assignment of authority shall be adequately documented and the FS shall be notified. The FS shall approve agreements involving any interests in lands held by the Federal

Government prior to such delegation or assignment. Once interests in lands are acquired, the State lead agency, FS, and others as appropriate, may negotiate tract-specific Memorandums of Understanding (MOU) as necessary to specify management and monitoring responsibilities for the interests in lands.

Optimal management and monitoring of tracts in FLAs is based upon partnerships between landowners, private non-profit organizations owning or managing lands, and State and Federal officials. Land trusts and other private organizations will continue to manage and monitor their own easements and lands within designated FLAs, and while they may not manage government-owned interests in lands under the FLP, they may cooperate with or contract for monitoring and implement specific management activities. Management of federally owned interests in lands is reserved to the FS, but may be assigned to State or local governments, or another Federal agency through mutual agreement. Although delegable, enforcement actions for easements will generally be conducted by the easement holder, i.e., the State or the Federal Government.

Monitoring FLP conservation easements shall occur periodically, but not less than annually. Monitoring consists of visual inspection of the property, documented by a written report to explain the condition of the property at time of inspection. Any material departure from the baseline documentation report or Forest Stewardship Plan should be noted.

The easement holder should immediately address any violation of the conservation easement with the landowner. The landowner should have the opportunity to correct the breach. After a reasonable time period (e.g. 30 days), if the breach is not corrected, enforcement action may be taken, including but not limited to, legal means.

The unit of government holding the conservation easement has the initial responsibility to enforce the conservation easement. The State or easement holder shall promptly notify any future FLP tract owner of the FLP and the origin and requirements of the conservation easement. The Forest Stewardship Plans covering the tract shall be reviewed periodically and updated as needed. If there is a change in land ownership, then the Forest Stewardship Plan needs to be reviewed, and updated as needed. -

Landowner Participation

Landowner participation in the program is voluntary and consists of two elements:

1. Conveyance of lands and interests in lands to achieve the purpose of the FLP;
2. Preparation and periodic updates of a Forest Stewardship Plan or a multi-resource management plan. The landowner and the State Forester or designee must approve the plan prior to signing the acquisition of the easement. The plan shall include provisions to meet land conservation objectives of the FLP. The plan shall be kept current and updated as needed. Modifications of the plan must be agreed to by the State lead agency. A plan is not needed if the lands are purchased in fee. (See Appendix F for sample content of a Forest Stewardship Plan) Landowners may submit an application and property information (See Appendix E) to the State lead agency to enroll their land or interests in lands in the FLP according to the process described in the AON. All owners of eligible forestlands within the designated FLA, and meeting the minimum Eligibility Criteria or other application requirements described in the AON, are eligible to submit an application.

For a landowner to participate in the program, it is not required that their tracts be completely forested. (see definition of "Nonforest Uses" and "Reserved Areas") However, priority will generally be given to tracts that are currently forested or are identified to be forested in the landowner Forest Stewardship Plan or multi-resource management plan.

The FLP respects the rights of private property holders. Under no circumstances shall the right of eminent domain be used for the unwilling "taking" of any private property rights. Traditional forest uses such as forest management activities, including timber management, and outdoor recreation opportunities are deemed consistent with purposes of the FLP and are encouraged on FLP tracts when consistent with the State's AON and the conservation purposes for FLP tract acquisition.

The FLP adheres to language contained in Section 14 of the CFAA, Statement of Limitation: "This Act shall not authorize the Federal Government to regulate the use of private land or to deprive owners of land of their rights to property or to income from the sale of property, unless such property rights are voluntarily conveyed or limited by contract or other agreement.

This Act does not diminish in any way the rights and responsibilities of the States and political subdivisions of States." Purchase or donation of rights does not relieve landowners of regulations that would otherwise apply.

The FS has no jurisdiction to make tax determinations or render advice as to the tax implications of transactions. Since tax implications differ from person to person, landowners should be encouraged to seek independent counsel from local assessors, tax lawyers, or accountants.

Land Trust Participation

Land trusts are nonprofit organizations that protect land by working with landowners wishing to donate or sell fee title or conservation easements to maintain conservation values associated with the land. Land trusts can have an important role to play in the FLP. The following considerations apply to land transactions between the Federal Government/States and land trusts:

1. Land trusts cannot execute contracts for acquisition of interests in lands on behalf of the Federal /State Government. Land trusts may work as intermediaries for eventual Federal/State acquisition, but without an accepted land purchase option and contract with the FS there is no guarantee of Federal acquisition. No pass-through transactions shall be done without prior consultation with the FS/State.

2. With approval of the State lead agency, the FS, the land trust or the donating landowner, lands and interests in lands acquired by land trusts (pursuant to Final Guidelines Part 1, Section XIII) may be counted toward the nonfederal cost-share contribution, provided that the interests in lands permanently contribute to the FLP.
3. If a land trust proposes a pass-through transaction to the FLP it must assure that terms and conditions in the deed or conservation easement are reviewed and approved in advance by the State lead agency and/or the FS.
4. The monitoring of easements within FLAs may be performed by land trusts in accordance with the umbrella MOU for the FLP in that State and individual MOUs for specific tracts established between the State and the land trust organization.
5. Other appropriate and beneficial roles of land trusts in relation to the FLP may include: participation on the SFSCC; recruitment and facilitation of FLP projects; buyer of tracts or easements of proposed, but unfunded FLP projects; facilitators of local FLP efforts; and performing tract monitoring and management activities.

PART 2-STATE GRANT PROGRAM

The State lead agency elects the State grant option of the FLP, in writing, to the appropriate FS Region/Area/IITF.

When a State elects the State grant option, all FLP acquisitions shall be transacted by the State with title vested in the State or a unit of State or local government. There are two exceptions:

1. Donations where the donor may wish to make a donation to a land trust, local, or Federal Government and the donee agrees to accept the donation, and to manage the lands or interests in lands in perpetuity for FLP purposes; and
2. At the request of the State and at the discretion of the FS, the FS may acquire individual tracts or multiple tracts within a specified FLA, with title vested in the U.S. in accordance with Part 3 of these guidelines.

Grants

If a State elects the optional State grant option, the FS will provide a Federal grant to the State to carry out the FLP, including the acquisition by the State of lands and interests in lands. Grants must be consistent with the uniform administrative requirements established in 7 CFR 3016.

States will generally be reimbursed for costs incurred with cash advances limited to the minimum amounts needed and timed to be in accord only with the actual, immediate cash requirements of the State in carrying out the FLP. The timing and amount of cash advances shall be as close as is administratively feasible to the actual cash outlay by the State for direct program costs and the proportionate share of any allowable indirect costs.

Conditions of the Grant

1. States must submit annual performance and financial status reports. A final performance report and financial status report are required prior to close out of the grant.
2. Funds appropriated for the FLP shall not be included in consolidated-payment grants made under authority of Section 12 of the CFAA.
3. The State shall maintain current and complete financial records in accordance with requirements contained in the latest Federal Aid Manual and OMB Circular.

Eligible Activities

The following activities are eligible uses of funds granted to States for the FLP; however, in most cases costs incurred prior to issuance of the grant cannot be reimbursed:

1. Purchase of lands or interests in lands from willing sellers for inclusion in the FLP;
2. Facilitation of donations of lands or interests in lands to a qualified and willing donee for FLP purposes;
3. Program administration expenses limited to indirect costs and direct acquisition related expenses for lands and interests in lands acquired under Forest Legacy authority;

4. Establishment and documentation of baseline conditions and development of a Forest Stewardship Plan for a conservation easement; and
5. AON Planning and amendment.

The following uses of Forest Legacy funds are not allowed as part of a State grant:

1. Management of acquired lands or interests in lands including, monitoring of conservation easements,
2. Enforcement actions, and
3. Payment for appraisals of donated property when the donation represents the full and total value.

Availability of Funds

Project funds for any fiscal year shall be available to the State for two years from the time they are obligated in a FS grant to the State in order to insure that Federal funds are spent promptly to acquire FLP projects. However, a grant may have a maximum duration of five (5) years to allow for nonfederal cost sharing to occur. During the 5-year life of the grant, it can be amended annually, as needed, and funds from a new fiscal year added to the grant, consistent with the requirement that the funds be expended within two years of the time of obligation. In no case can funds be obligated or expended beyond the 5-year life of the grant.

Acquisition of Lands and Interests in Lands by States

All Forest Legacy acquisitions including the acquisition of lands or interests in land shall be made in accordance with Federal appraisal and acquisition standards and procedures. The interests in land acquired for Forest Legacy shall be adequate for Forest Legacy purposes and be perpetual. Title to such lands or interests in lands will be vested in the State or unit of State government. These lands or interests in lands will be managed and administered for goals consistent with Forest Legacy conservation purposes by State agencies or their assigns. The State agencies are responsible for all monitoring and management of conservation easements and management of fee simple properties.

Lands and interests in land located within a FLA and simultaneously within other Federal boundaries (e.g. national forest, national park, or national wildlife refuge) are eligible for the FLP provided that the responsible Federal agency concurs with the FLP State acquisition. If a State has passed legislation that extinguishes claims to or restrictions on real property, the State shall use all available authorities, including that of acting as an agent of the U.S., to achieve the purposes of section 7(K)(2) of the CFAA.

Reversion of Funds for Forest Legacy Inconsistency

In the event it is determined, by the State lead agency, that it is no longer desirable to hold lands or interests in lands acquired with Federal funding and those lands are conveyed, exchanged, or otherwise disposed of, after providing notice to the FS, the State shall:

1. Reimburse the FS for the current market value in proportion to the original Federal investment; (said reimbursements to be used to further the purposes of the FLP); or
2. Exchange for other FLP eligible lands or interests in lands of at least equal market value and of reasonably equivalent location, with public purposes that equal or exceed those of the disposed tract, with FS approval.

Items 1 and 2 identified above must be included in deeds or conservation easements of all FLP tracts as well as in the FS grant to the State. Appendix I includes suggested language for conservation easements and deeds.

PART 3-FEDERAL ACQUISITION PROGRAM GUIDELINES

Federal Acquisition Process

In the furtherance of the purposes of the FLP, the State lead agency with involvement of the SFSCC and the FS will review property owner applications, prioritize tracts, obtain State approval, and submit properties to the appropriate FS Region/Area/IITF for funding. Upon approval for funding, the FS will proceed to acquire from willing sellers conservation easements and/or other interests in land including fee acquisition. Federal Acquisition Procedures must be followed when Federal funds are used to complete an acquisition of land or interests in land using FLP authority. They are:

1. Federal appraisal standards must be met;
2. The landowner must be informed of the market value and that sale of the property is strictly voluntary;
3. The landowner must be notified in writing that the property will NOT be purchased if negotiations do not result in an amicable agreement;
4. Federal payment to the landowner for lands or interests in lands is not more than the market value determined under #1;
5. Assure title is free and unencumbered relative to the purposes of the FLP; and
6. If relocation is involved the requirements in PL 91-646 (42 U.S.C. 4601) must be followed and the FS must advise the landowner prior to the acquisition.

Certain lands are not eligible for the Federal ownership option under FLP authority because other authorities and funding sources are available for acquisition of lands or interests in lands within these federally established areas. These include lands or interests in lands located within National Forests, National Parks, National Wildlife Refuges, or other Federal Government boundaries. Proximity to Federal lands or the inclusion of Federal lands within a proposed FLA does not disqualify an area for program eligibility.

Federal laws governing public lands do not apply to private property rights not acquired by the Federal Government from willing private landowners. Interests in lands retained by private landowners, not conveyed to the Federal Government under the FLP, are subject to the same requirements of the Endangered Species Act (ESA) that existed prior to their participation in the FLP. Conveyance of interests in lands to the Federal Government neither enhances nor diminishes the landowner's responsibility under the ESA. Any interests in lands acquired by the Federal

Government under the FLP shall be subject to the same requirements of the ESA as are other Federal lands.

Memorandum of Understanding (MOU) for Coordination of the FLP

An MOU will be used to coordinate the FLP where Federal acquisition option resulting in Federal ownership of FLP acquisitions occurs. The MOU will define and facilitate partnerships between the State lead agency, FS, and other participating entities in implementing the program, acquiring interests in lands, and sharing the costs of the program. The MOU shall determine how costs are shared between parties, including administrative, management, monitoring, and capital improvement expenses. The terms of a MOU will determine which party is responsible for costs incurred following the project's five-year cost-share write off period. If individual Forest Legacy tract-MOUs are needed, they become an addendum to the State level "umbrella" MOU. The umbrella MOU between the State lead agency and the FS shall be developed following the Secretary's approval of the State's AON and the establishment of the State's FLP.

The FS/State MOU is for the purpose of specifying roles and responsibilities for implementing the program, and may address the following items:

1. Costs and Funding:
   a. Identify direct and indirect costs expected to be incurred in establishing the FLP, and acquiring and administering interests in lands during the first five years of the program. Revise or renew these cost estimates as appropriate.
   b. Identify and propose sources of cost-share matches.
2. Planning:
   a. Document the amount of work required to complete the AON and identification of FLAs.
   b. Define a process for revising existing landowner Forest Stewardship Plans, or multi-resource forest management plans.

c. Identify how specific tract acquisition needs and priorities shall be established by the State.

3. Acquisition:

   a. Identify who is responsible for title work, appraisals, surveys, and similar pre-acquisition work.

   b. Define a process for determining the value of donated interests in lands.

4. Management:

   a. Define responsibilities for management of interests in lands acquired or dedicated to the program.

   b. Identify possible activities needed to enhance, restore, or maintain resources to meet the intent of the program and general responsibilities in carrying out such activities.

5. Administration:

   a. Estimate the staff-work required to implement the Program.

   b. Define responsibilities for processing applications to the FLP.

   c. Establish procedures for monitoring and enforcement the terms of reserved interest deeds and easements and identify who will be responsible.

   d. Identify responsibilities for periodic reports summarizing the achievement of FLP goals in the State.

Payment in Lieu of Taxes (PILT)

Where the Federal Government under the FLP acquires lands in fee, the Federal Government will pay PILT to the local taxing authority. No PILT will be paid on conservation easements.

Transition to State Grant Option Program

If a State elects the State Grant Option, and there are active cases being pursued by the FS, all parties (FS, State, and landowner) may agree to transfer the case to the State. If agreement to transfer is reached, then the value of the lands or interests in lands comprising the project may be transferred to the State by a FS grant. To facilitate projects transferred to the State, the FS may provide the State with copies of any appraisals, appraisal reviews, title reports, option contracts and other pre-acquisition materials for lands that have been under negotiation by the FS.

EXAMPLE OF A PROJECT SELECTION CALENDAR

This flowchart outlines the basic FLP project selection process.

JANUARY
2003
FEBURARY
MARCH
APRIL
MAY
JUNE
JULY
AUGUST
SEPTEMBER
OCTOBER
NOVEMBER
DECEMBER
JANUARY
2004
FEBURARY
MARCH
APRIL
MAY
JUNE
JULY
AUGUST
SEPTEMBER
OCTOBER

FS completes FY 2005 Recommended Project Briefing Book State Lead Agencies receive applications for FY 2005 FLP projects (some States have an open application process) State Forest Stewardship Coordinating Committees meet to evaluate and approve applications for FY 2005 funding according to the criteria identified in the State AON.

FS regional units consult with States to create a FY 2005 "Project Opportunity List" that includes all projects that they are requesting funding for in priority order.

FS regional units develop a FY 2005 "Project Recommendation List" based on regional criteria, State project lists and project briefings and submit to FS Washington Office.

FS regional units submit FY 2005 "Project Opportunity List" to FS Washington Office.

FS conducts National Review Panel representing FS and State regions to prepare a prioritized FY 2005 "Forest Legacy Project.

Recommendation List" for submission to Administration.

FS submits FY 2005 FLP Project Recommendation List to the Administration.

FS notifies the House and Senate Appropriations Committee on the project list after it clears the Administration.

Congressional committees discuss and pass annual Appropriations for FY2005 projects.

APPENDIX C-OFFICE OF MANAGEMENT AND BUDGET (OMB) CIRCULARS AND OTHER REGULATIONS

Any award of Federal financial assistance under these guidelines will be subject to the following or its most recent update:

1. OMB Circular A-102 (10/7/1994, amended 8/29/1997), "Grants and Cooperative Agreements with State and Local Governments"
2. OMB Circular A-87 (5/4/1995, amended 8/29/1997), "Cost Principles for State, Local, and Tribal Governments" as implemented by Departmental Regulation 7 CFR 3016,

"Uniform Administrative Requirements for Grants and Cooperative Agreements to State and Local Governments"

3. OMB Circular A-110 (11/19/1993, amended 09/30/1999), "Uniform Administrative Requirements for Grants and Other Agreements with Institutions of Higher Education, Hospitals and Other Non-Profit Organizations"
4. OMB Circular A-122 (6/1/1998), "Cost Principles for Non-Profit Organizations"
5. OMB Circular A-133 (06/24/1997), "Audits of States, Local Governments, and Non-Profit Organizations" as implemented by Departmental Regulation 7 CFR 3050, "Audits of State and Local Governments" OMB Circular A-89 (8/17/1984), "Catalog of Federal Domestic Assistance"
6. 7 CFR 3017, Government Debarment and Suspension (Nonprocurement) and Government-wide Requirements for Drug-Free Workplace (Grants), and
7. 7 CFR 3018, New Restrictions on Lobbying.
8. 7 CFR 3019, Uniform administration requirements (Higher education, hospitals, and non-profit organizations"

EXAMPLES OF COST SHARE CALCULATIONS

Equation for Calculating Cost Share Requirement (Federal FLP Share) X (0.333) = the minimum Non-Federal Contribution

OR

(Total Project Costs) X (0.75) = the maximum Federal Contribution

Principals to Guide Calculating the Cost-Share Requirements

- o To calculate the cost share requirement, the Program Manager should use the Federal FLP contribution, and not the total project costs.
- o The cost share requirement should be at least 33.3% of the total Federal FLP contribution towards the project, which will equal at least 25% of the total FLP project (Federal FLP contribution plus cost share).

- o The Federal contribution (Forest Service's FLP plus all other Federal contributions) cannot exceed 75% of the total project costs (all cost requirements to complete the project, including Federal and non-Federal contributions).
- o The non-Federal cost share portion cannot be used as cost share for another Federal program that also requires a cost share.

Example 1-The FLP is going protect Jane Smith's 3,000 acres tract. The total cost of protecting that land is $1 million.

Total Project Costs	Federal FLP Contribution	Non-Federal FLP Contribution	Other Federal Contribution	Other non-Federal Contribution
$1,000,000	$750,000	$250,000	$0	$0

- o Federal contribution cannot exceed $750,000; therefore, the Federal contribution is not greater than 75% of the total project costs.
- o The non-Federal cost share requirement is at least $250,000; therefore, FLP funds are adequately cost shared.

Example 2-John Doe Ranch is planning to conserve 6,500 acres of land. The total cost of protecting the land is $4 million. The Federal contribution, through FLP, will be $1,000,000, and the non-Federal contributors will provide $3,000,000, which includes a cost-share component for the FLP.

Total Project Costs	Federal FLP Contribution	Non-Federal FLP Contribution	Other Federal Contribution	Other non-Federal Contribution
$4,000,000	$1,000,000	$333,000	$0	$2,667,000

- o Federal contribution cannot exceed $3,000,000; therefore, the Federal contribution is not greater than 75% of the total project costs.
- o The non-Federal cost share requirement is at least $333,000; therefore, FLP funds are adequately cost shared.

Example 3-ABC Tree Company is planning to conserve 8,300 acres of land. Both theForest Service's FLP and the U.S. Fish and

Wildlife Service (FWS) are contributing funds toward the project. Non-Federal money has been secured to cover the non-Federal cost share requirements for the FLP and FWS requirements, as well as to pay for additional project costs.

Total Project Costs	Federal FLP Contribution	Non-Federal FLP Contribution	Other Federal Contribution	Other non-Federal Contribution
$4,000,000	$1,000,000	$333,000	$1,000,000	$1,667,000

- o Federal contribution cannot exceed$3,000,000; therefore, the Federal contribution is not greater than 75% of the total project costs.
- o The non-Federal cost share requirement is at least $333,000; therefore, FLP funds are adequately cost shared.
- o FLP cost share component cannot be the same as the FWS cost share component.

INFORMATION TO FACILITATE LANDOWNER PARTICIPATION

Landowners who wish to participate in the program may be asked to provide the following information.

1. Name, address and phone number of applicant landowner.
2. All other owners of record for this tract, and their addresses.
3. Name, address and phone number of authorized agent representing landowner(s) if applicable.
4. Location of property.
5. If the landowner intends to reserve rights to forestry uses or other resource management activities, a copy or reference to the State approved landowner Forest Stewardship Plan or multi-resource management plan.
6. List of the significant scenic, natural, recreational, wildlife, timber and other resource values contained on the property.
7. Identification of all dams, dumps or waste disposal sites on the property.
8. Signed statement giving the FS and State lead agency permission to enter the property for review and appraisal purposes.

9. Legal description.
10. List any encumbrances or liens existing on the property including, but not limited to contracts, leases, or outstanding rights not of record.
11. Copy of plat or survey map of the property, if existing. If only a portion of the property is being offered, identify it on a plat showing the portion offered in the context of the entire tract.
12. Tract acreage and total number of acres of forests and cleared/open land.
13. List of existing permanent improvements on the tract, including houses, barns, lakes, ponds, dams, wells, roads, and other structures, and total number of acres occupied by improvements.

SAMPLE CONTENT OF A FOREST STEWARDSHIP PLAN

Below is information from the Forest Stewardship Program's National Standards and Guidelines. Please also refer to the Forest Stewardship Program's Planning for Forest Stewardship: A Desk Guide as well as States' Statewide Forest Stewardship Plans for additional information on Forest Stewardship Plans.

Landowner Forest Stewardship Plans must:

- o be prepared or verified, as meeting the minimum standards of a forest stewardship plan, by a professional resource manager.
- o identify and describe actions to protect, manage, maintain and enhance relevant resources listed in the law (soil, water, range, aesthetic quality, recreation, timber, water, and fish and wildlife) in a manner compatible with landowner objectives.
- o be approved by the State Forester or a representative of the State Forester.
- o involve the landowner in the plan development by setting clear objectives and should understand clearly the completed plan.

A well prepared plan will:

- o Clearly state landowner objectives.
- o Have a cover page.
- o Provide for authorship and/or signature lines within the document.

The plan preparer should consider and evaluate resource elements present and include a brief description of those that are applicable and their importance to the ownership. Resource elements to be considered are:

- o Soil Interpretations
- o Water
- o Range
- o Aesthetic Quality
- o Recreation
- o Timber
- o Fish
- o Wildlife
- o Forest Health
- o Archeological, Cultural and Historical Sites
- o Wetlands
- o Threatened and Endangered Species

Management recommendations, or where appropriate, alternative strategies should be provided for those resource elements described. Prescriptions or treatments should be integrated and stand or site specific. An ownership map drawn to scale, or photo, to include vegetation cover types, stream and pond location with a legend will enable the landowner to implement the plan. Landowners' understanding may be improved by including activity summaries and appendices.

Appendices might include:

- o Description of assistance available and incentive programs
- o Educational materials

- o A glossary of terms
- o An explanation of applicable Federal, State and /or county regulatory programs, especially as they apply to:
- o Archeological, cultural and historical sites.
- o Wetlands.
- o Threatened and Endangered Species.

These last three items are covered by legislation other than the Cooperative Forestry Assistance Act of 1978, as amended by title XII of the Food, Agriculture, Conservation and Trade Act of 1990 (16 U.S.C. 2101, et seq.), but must be considered for Federally funded programs. The professional resource manager should discuss the Forest Stewardship Plan with the landowner, following completion, to assure understanding.

REAL ESTATE RECORD KEEPING

Since Forest Legacy acquisitions are perpetual, record keeping is important. Each State shall maintain permanent records for all Forest Legacy properties. The following information is recommended to be maintained by the conservation easement holder:

A. Landowner information (name, address, phone)
B. Nomination form (including notification to landowner that property will not be purchased if negotiations do not result in amicable agreement)
C. Landowner Inspection Consent Agreement
D. Baseline documentation
E. Option agreement
F. Deed of Conservation Agreement
G. Additional warranty deeds, covenants, restrictions
H. Title Insurance Policy
I. Appraisal
J. Appraisal review
K. Forest Stewardship Plan or equivalent

L. Notification of county or local government

M. Closing statement

N. Copies of check or documentation of EFT or other form of payment

O. Copies of grant reimbursement or expenditure

The following items should also be maintained as part of the record:

1. Landowner correspondence
2. Evaluation criteria
3. Tracking/documentation of negotiation steps
4. State Forest Stewardship Coordinating Committee recommendation
5. Press release
6. Monitoring records/history

REQUIRED QUALIFICATIONS OF AN APPRAISER OR REVIEW APPRAISER

A. Appraiser-In order to be a qualified appraiser for purposes of FLP appraisals, an individual must be:

1. a Federal land acquisition agency staff appraiser who
    a. is certified as a general appraiser in compliance with OMB Bulletin 92-06, and
    b. has completed training in application of the December 2000 edition of Uniform Appraisal Standards for Federal Land Acquisitions(UASFLA)* approved for appraiser continuing education credit in the State where the appraiser is certified, or
2. a nonfederal staff or fee appraiser who
    a. is certified as a general appraiser in the state where the appraised property is located, or can obtain reciprocity or a temporary practice permit in the state where the appraised property is located, and

b. has, within the past 10 years, completed at least the minimum classroom hours of non-duplicative education prescribed for the certified general real property appraiser classification by the Appraisal Standards Board of The Appraisal Foundation, and

c. has completed at least 12 self-contained or summary appraisal reports of properties similar in scope and complexity to the appraised property in the preceding three years, and

d. has completed training in application of the December 2000 edition of Uniform Appraisal Standards for Federal Land Acquisitions approved for appraiser continuing education credit in the state where the appraiser is certified.

The qualified appraiser shall prepare an appraisal report in compliance with the Uniform Appraisal Standards for Federal Land Acquisitions and supplemental written appraisal instructions issued by the client. Federal land acquisition agencies are the member agencies of the Interagency Land Acquisition Conference.

B. Review Appraiser-In order to be a qualified review appraiser for purposes of FLP appraisals, an individual must be:

1. a Federal land acquisition agency staff appraiser who
   a. is certified as a general appraiser in compliance with OMB Bulletin 92-06, and
   b. holds specific delegated authority to review and approve or recommend appraisals for agency use, and
   c. has completed training in application of the December 2000 edition of UASFLA* approved for appraiser continuing education credit in the State where the reviewer is certified, or

2. a nonfederal staff or fee appraiser who:
   a. is certified as a general appraiser in the State where the appraised property is located, or can obtain reciprocity or a temporary practice permit in the state where the appraised property is located, and
   b. has, within the past 10 years, completed at least the minimum classroom hours of non-duplicative education prescribed for the certified general real property appraiser classification by the Appraisal Standards Board of The Appraisal Foundation and at least 32 classroom hours of approved training in appraisal review, or otherwise demonstrates competency in appraisal review in compliance with the Competency Rule of the Uniform Standards of Professional Appraisal Practice (USPAP), and
   c. has completed at least 12 self-contained or summary appraisal reports of properties similar to the appraised property in the preceding three years or at least 12 technical appraisal review reports for appraisal reports of properties similar in scope and complexity to the appraised property in the preceding three years, and
   d. has completed training in application of the December 2000 edition of Uniform Appraisal Standards for Federal Land Acquisitions approved for appraiser continuing education credit in the state where the reviewer is certified.

The qualified review appraiser shall prepare a technical appraisal review report that includes a determination of whether the appraisal report under review complies with the Uniform Appraisal Standards for Federal Land Acquisitions. Federal land acquisition agencies are the member agencies of the Interagency Land Acquisition Conference. *The seminar, Federal Land Exchanges and Acquisitions: Appraisal Issues and Applications, offered by the American Society of Farm Managers and Rural Appraisers and the Appraisal Institute is the only acceptable substitute for UASFLA training.

REQUIREMENTS AND SUGGESTIONS FOR CONSERVATION EASEMENTS AND DEEDS

The Purpose and Authority and Reversion clauses are required in all FLP easements and deeds. Below are examples of language that States have used to meet that requirement:

Purpose and Authority Clause

Example 1

WHEREAS, the Conservation values of the Property are consistent with the goals of the Forest Legacy Program and the establishment of this conservation easement will provide public benefits by:

> *preventing future conversions of forest land and forest resources; protecting and enhancing water quality and water supplies; protecting wildlife habitat and maintaining habitat connectivity and related values to ensure biodiversity; protecting riparian area; maintaining and restoring natural ecosystem functions; and maintaining forest sustainability and the cultural and economic vitality of rural communities.*

WHEREAS, the specific Conservation Values of the Property are documented in an inventory of relevant features of the Property. The data and explanatory text are presented in the Baseline Documentation Report, dates ______, which consists of reports, maps, photographs, and other documentation that the parties agree to provide.

This Easement acquisition is authorized by the Cooperative Forestry Assistance Act of 1978, as amended by section 1217 of the Food, Agricultural, Conservation and Trade Act of 1990 (104 Stat. 3528; 16 U.S.C. Section 2103c).

Example 2

The purpose of this easement is to effect the Forest Legacy Program in accordance with the provisions of Title XII of the Food,

Agriculture, Conservation and Trade Act of 1990 (16 U.S.C.-2103c) as amended, on the herein described land, which purposes include protecting environmentally important forest areas that are threatened by conversion to non-forest uses and for promoting forest land protection and other conservation opportunities. The purposes also include the protection and preservation of important scenic, cultural, fish, wildlife and recreational resources, riparian areas, and other ecological values, and to ensure that the Property is available for the sustainable and cost effective harvesting of forest products in a silviculturally sound manner, all of which meet the objectives of the Forest Legacy Program. The purposes also include encouragement of management for and the production of economically sustainable and commercially viable forest products consistent with the other purposes of this easement and also include the long-term protection of the Conservation Property's capacity to produce economically valuable forestry products, and the encouragement of management of the property for industrial or commercial forestry only if consistent with the other purposes of this Conservation Easement.

The Parties agree that the purpose of this easement is also to assure that the Property herein described as Schedule "A" and hereby encumbered as set forth in Schedule "B" will be retained forever in its existing natural, scenic and forested condition and to prevent any use of the Property that will significantly impair or interfere with the conservation values of the Property. The Grantor intends that this easement will confine the use of the Property to such activities specifically enumerated herein which are consistent with the overall purposes of the easement by protecting the following particular values of the easement area: specifically the scenic, cultural, fish, wildlife and recreational resources, riparian areas and similar ecological values.

Example 3

WHEREAS, the clearly delineated open space conservation goals and objectives as stated in Forest Legacy Program pursuant to Section 1217 of Title XII of the Food, Agriculture, Conservation and Trade Act of 1990 (16 USC Section 2103C) which was created "to protect environmentally important private forest lands threatened with conversion to non-forest uses" has awarded a Forest Legacy grant in to the Grantors for purchase of a portion of the value of the Easement herein conveyed for a conservation easement on forestal, agricultural, and open space land.

Example 4

The purpose of this easement is to effect the Forest Legacy Program in accordance with the provisions of Title XII of the Food, Agriculture, Conservation and Trade Act of 1990 (16 U.S.C. 2103c) on the herein described land, which purposes include protecting environmentally important forest areas that are threatened by conversion to nonforest uses and for promoting forest land protection and other conservation opportunities. The purposes also include the protection of important scenic, cultural, fish, wildlife and recreational resources, riparian area, and other ecological values.

Example 5

The purpose of this conservation easement is to restrict the exercise of all development rights, residential, commercial or otherwise, on the easement area and to protect the scenic and recreational values of said easement area from conversion to non-forest uses while at the same time allowing for the use of the area for commercial forestry and public recreation purposes consistent with the stated purposes, standards and general intent expressed in Title XII of the Food, Agriculture, Conservation and Trade Act of 1990 (16 USC 2103c) and the requirements of Section 7 for the Forest Legacy Program.

Reversion Clause

The Easement Holder acknowledges that this Easement was acquired with Federal funds under the Forest Legacy Program (P.L. 101-624; 104 Stat. 3359) and that the interest acquired cannot be sold, exchanged, or otherwise disposed, except as provided in Section 5.A, unless the United States is reimbursed the market value of the interest in land at the time of disposal. Provided, however, the Secretary of Agriculture may exercise discretion to consent to such sale, exchange, or disposition upon the State's tender of equal valued consideration acceptable to the Secretary.

SAMPLE CONTENT FOR BASELINE DOCUMENTATION

The following list has been modified from the Checklist included in the Land Trust Alliance and Trust for Public Land's The Conservation Easement Handbook (1988).

1. Cover Page
   - o including name and location of property, signature of the author/collector, and date
2. Table of Contents
3. Owner Acknowledgement of Condition.
4. Background Information
   - o Ownership information (name, address, and phone number of property owner)
   - o Historical information on the donation/acquisition (brief chronological description of events that led to the protection of the property)
   - o Summary of easement provisions (specific prohibitions, restrictions, retained rights, as derived from the language of the easement document)
   - o Purpose of easement
   - o Evidence of the significance of the protected property, as established either by the government policy (include copies of documents) or by the long-term protection strategy developed by the grantee
   - o Corporate or agency resolution accepting gift (minutes of the meeting at which a gift is accepted or acquisition approved are adequate)
5. Legal Condition
   - o A copy of the signed, recorded easement document
   - o An assessor's parcel map

- o A clear title statement or preliminary title report, noting any liens against the property that could compromise its natural qualities or invalidate the easement
- o Copies of any other relevant easements or water rights associated with the property

6. Ecological Features
   - o A general description of the ecological features that the easement seeks to protect, such as forest and plant communities, soil characteristics, and habitat.
   - o The Forest Stewardship Plan should be used as a guide to determine what information is needed.
   - o An inventory of rare, endangered, and/or threatened species and habitat found on the property
   - o Reports from wildlife biologists or other specialists that document the status of significant natural elements
7. Agricultural Features
   - o Intensity of grazing (can be determined by experts and expressed in "animal units" per acre) and farming
   - o Level of pesticide use
8. Scenic Features
   - o Official policies citing property's scenic value
   - o Number of people who frequent nearby public places (roads, trails, parks) from which they can view property
9. Archeological, Cultural and Historical Features
   - o Archeological, cultural and historical sites and resources found within the property, with a focus on those resources that the easement seeks to protect.
10. Human Created Features
    - o Improvements (structures, trails, fences, wells, power lines, pipelines, irrigation systems, etc.)
    - o Recreation/tourism attractions
    - o Trespass damage and disturbed land (stray animals, introduced species evidence of vehicular trespass, etc.)

11. Photographs
    - o Aerial photos, if appropriate
    - o On-site photos (be sure to record key photo points, record distance and azimuth from structures or other fixed points, and sign and date all photos)
12. Maps
    - o A state map showing easement location
    - o An 8 ½" X 11" section of a local road map showing easement location
    - o The largest scale U.S. Geological Survey topographical map available (usually at a scale of 1:24,000, called a 7-1/2 minute scale), showing easement boundaries
13. Survey
    - o Surveys generally are not required, but may be helpful

3

National Forestry Action Plan

This chapter discusses about the findings and recommendations of a Country Mission to Antigua & Barbuda which was conducted 13 October to 23 November, 1991. The mission was a fifth in a series being undertaken for the Tropical Forests Action Programme (TFAP) in nine countries of the eastern Caribbean Community (CARICOM).

In Antigua and Barbuda, the mission was supported by the German Institute for Technical Cooperation (GTZ), ODA-UK and FAO in recruiting a multi-disciplinary team of eight international consultants covering a range of specialities.

The aims and objectives of the mission were to identify the principal causes of deforestation and to facilitate preparation of a 20-year forestry development plan which would ensure that:

(i) appropriate land use policies are put in place to protect and conserve the national forest estate;

(ii) all existing natural forests are managed in perpetuity on a sustained yield basis in accordance with sound environmental management techniques;

(iii) optimal use is made of reforestation and agroforestry techniques to provide environmental protection where required on both State and private lands and to assist in the rural development process;

(iv) high priority is given to the identification and legal protection of special biological interest, through the establishment of legally constituted Nature Reserves;

(v) primacy of forest conservation, sustained yield management and protection of the environment are the overwhelming responsibility of the forest manager in seeking to optimize the utilization of products from the forest;

(vi) provision is made for the protection of wildlife and for forest recreation facilities in the form of legally constituted National Parks.

The exercise was carried out with the close collaboration and active participation of national counterparts from relevant public agencies and representatives of the private sector and NGOs. Working papers were prepared for circulation and for use as reference documents. These papers were used by the author in synthesising the main report in which the situation of Antigua and Barbuda was analyzed according to the five TFAP priority areas (see regional overview document). Based on these analyses, and group discussions of the major issues during the mission, strategies and action programmes were developed to attract technical assistance and investment with the aim of optimising the contribution which forestry and tree cover can make to sustainable social and economic development of the country.

The report, therefore, represents a National Forestry Action Plan (NFAP) for Antigua and Barbuda. The NFAP recognises the limitations of the country's economy but stresses the long-term benefits of environmental management and development. The Government endorsed the proposals before they were presented to the public for discussion at a National Round Table held in June, 1991. The active and meaningful feedback generated allowed the draft NFAP document to be strengthened and made available to the international community for consideration of technical and financial support. It is hoped that this International Round Table will be convened towards the latter half of 1993.

Reference should also be made to related documents:

Regional Overview of the CARICOM/FAO Tropical Forests Action Programme in the East Caribbean. CARICOM/FAO/BDDC-ODA, July 1993.

Forestry development initiatives at the regional/subregional level-regional project proposals 01-05.

The Status of Mangrove Conservation in the CARICOM Islands of the Eastern Caribbean. Report to the Commission of the European Communities as part of the TFAP for the Caribbean Region. P. Bacon, 1991.

BACKGROUND INFORMATION ON ANTIGUA AND BARBUDA

Natural Regions and Ecosystems

Antigua and Barbuda is located between latitude 17°0' and 17°45' N and longitude 62040' and 62°54' W, and has a total land area of 440 km 2 (Antigua 280 km2, Barbuda 160 km2) and population of 65,200 (Antigua 64,000, Barbuda 1,200).

The climate is tropical-maritime, with a mean annual temperature of 26°C, varying from 16.7 to 28.3°C. Average annual precipitation is 1100 mm, but variable throughout the year; the islands are susceptible to severe droughts (as well as earthquakes and hurricanes). The main rainy season is from October to November.

The geology of Antigua consists of a hilly limestone zone to the east, a central plain of marl and clay overlying volcanic material, and volcanic mountains in the southwest rising to 410 m above sea level at Boggy Peak. The topography is mainly gently undulating with slopes of 11-20 degrees. Shallow soils overly volcanic material and the limestone, over most of the island. Barbuda, which lies 32 miles north west of Antigua, is a low-lying coral limestone island, no higher than 60 m in the east, but with a large coastal lagoon enclosed by a long sand spit in the west.

Like most of the Caribbean Islands, Antigua was completely wooded when "discovered", but following European settlement it was almost totally cleared to make way for cultivation of sugar cane and cotton. With sugar as the dominant crop for over 300 years, Antigua & Barbuda's natural landscape was undergoing irreversible changes which created negative pressures on wildlife habitat and watershed areas. The demise of the sugar industry in the 1970's led to the abandonment of sugar cane plantations. Presently, Barbuda is affected by the overgrazing of animals which have continually removed valuable vegetation protecting watersheds.

As described by the Wildlife/National Parks consultant, the ecosystems of Antigua and Barbuda range from evergreen woodlands, xerophytic woodlands, scrublands, grasslands to mangrove forests, herbaceous swamps, saltponds, sandy beaches, rocky shores, coastal lagoons, sea grass beds, coral reefs, coral reef outcrops and oceanic islands and rocks. Ecosystem variety is enhanced by the presence of caves in the north-eastern limestone area and by natural seasonal drainage channels, streams, potholes and ponds. Barbuda's giant coastal lagoon is unique to the Region and together with its extensive tidal flats, sand bars, underwater sand dunes, salt ponds, cliffs, caves and "highlands", all providing special habitats for wildlife. Wildlife species associated with these natural ecosystems have developed physical, behavioural and physiological adaptations.

REVIEW OF THE FORESTRY SECTOR

With the decline in agriculture, the area of land under tree cover has increased over the years, and various reports estimate that there are 9,254 ha of woodland (33% of Antigua's land area) and 10,000 ha of scrub growth (36% of the land area). Table below classifies the woodland areas of Antigua.

Table : Woodland Composition for Antigua

WOODLAND	AREA (hectares)	WOODED AREA (percentage)
Moist Forest	2,200	24
Dry Woodland	5,654	61
Cactus Scrub	1,100	12
Mangroves	300	3
TOTAL	9,254	100

The species composition and structure of the vegetation has been described (Beard, 1949) and although all the forests are of secondary growth, they contain a number of valuable timber species, including mahogany (Swietenia mahagoni) and white cedar (Tabebuia pallida). On the present sites, however, they are unlikely to produce quality timber, as most of the forested lands are only suitable for conservation purposes.

Table : Woodland Distribution for Barbuda.

CATEGORY	AREA (hectares)	LAND AREA (percentage)
Wooded Areas	10,619	66
Mangrove Swamp	3,533	22
Scrub & Rough Grazing	1,285	8

Role of Forestry and Trees in the National Economy

As a small open market, forestry and related activities can play only a supplementary role in the Antigua and Barbuda economy which is primarily driven by the tourism industry, in the wake of the steady decline of agriculture.

Between 1977 and 1990, the country experienced strong economic growth, supporting one of the highest per capita standards of living among the OECS countries in the Eastern Caribbean. During this period, real GDP increased from EC $158.8 millions to EC $355.9 millions, an average growth rate of 6.2% per annum. However, the Government has been running a fiscal deficit for a number of years, having financed its capital investment programme through loans, a significant portion of which are denominated in foreign currencies. The Government has therefore announced austerity measures which include a freeze on public sector spending, more effective revenue collection and divestment of assets.

Nevertheless, tourism continues to grow, contributing an estimated 60% to the economy. However, it has placed considerable demands on the local infrastructure, particularly water and sewage, and crowded out alternative investment opportunities, while severely stressing the use of the natural environment for construction of additional resort facilities. The Forest Economics consultant has pointed out that changes in the market place can be affected by changes in societal values and a variety of products in the developed countries have suffered because they were determined as being "environmentally unfriendly". The tourist industry is dependent on visitors from the metropolitan countries where this discrimination is particularly rife. In these circumstances, it would be prudent for the Government to ensure that natural resources which enhance the tourism plant are well protected and, at the same time, properly utilised.

In this respect, the GOAB needs to adopt an enlightened approach to the planning, development and use of land and to ensure that such a policy is strictly observed. Under such a policy, forestry's role should be recognised as being essential to the protection and conservation of critical resources, such as water, and complementary to developing the tourism product, as well as the quality of life of the people of Antigua and Barbuda.

Other benefits which can accrue from forestry in Antigua and Barbuda include:

(i) rural employment opportunities;

(ii) shelter from the desiccating effects of the prevailing winds;

(iii) land stabilisation and soil conservation/amelioration;

(iv) production of fence posts, forage for livestock, fuelwood, fruits, nuts and other products;

(v) protection of natural habitats and wildlife for recreation, scientific research and tourism.

Despite the reputed benefits of the tourism industry, the Government should concentrate its attention on all opportunities for economic growth and optimal use of natural resources. In tandem with other national strategies, the role of forestry and trees needs to be heightened to become an integral contributor in maintaining living standards on a sustained basis.

Relevant Development Policies

In pursuing the development of the tourist sector, the GOAB appears to have neglected agriculture and the need to maintain stability in all sectors. Indeed, there seems to have been an absence of clear Government policies to guide the society's growth and improvement. However, it is pleasing to report that within the last two years a Ministry of Planning has been established in Antigua & Barbuda with the responsibility for developing a National Development Plan.

This new Ministry has initiated an integrated planning process which will take some time to fill the existing void. Such planning will take account of the country's present economic circumstances, the nature of its physical resources, its human resource capabilities and requirements in order to determine fundamental national strategies on which future growth can rely. In the short-term, the austerity measures announced must be equitably implemented as soon as possible, along with appropriate measures to ensure cohesion among the various sectors. The proposed NFAP recognizes these needs, and its multi-disciplinary approach is recommended for adoption by the Government in planning for the country's future.

Legislation

Antigua and Barbuda's forest legislation (The Forestry Act, Cap. 99) dates back to 1941 and provides for, inter alia, the establishment of reserves, exploitation permits, clearing and afforestation on private lands, etc. The Forest Regulations deal with permits, control of fires, cattle trespass and survey and demarcation of Forest Reserve boundaries. Unfortunately, the Act and Regulations are no longer enforced, and uncontrolled exploitation (mainly wattle for fish and lobster traps) has been practised. However, the forest cover is still relatively intact, but species diversity has been impaired.

The Legislation consultant has discussed related legislation and described various deficiencies existing in the enactments supporting resource and environmental management. Related legislation includes:

- The Land Development and Control Act (1977) for physical development planning;

- The Fisheries Act (1983) for protection of mangroves and habitats of fauna and flora;
- The National Parks Act (1984) to protect and develop ecological resources;
- Antigua Agricultural Development Corporation Act (1978) to stimulate agriculture (including forestry) and undertake commercial management.

The deficiencies identified underscore the need for new legislation in the forestry and water resources areas, together with amendments to other legislation, especially the National Parks Act, as well as the need for regulations to provide guidance for environmental managers. Amendment to the National Parks Act should include consolidation of existing laws relating to coastal zone management and declaration of protected areas, viz. Beach Protection Act (1957), Beach Control Act (1959), Marine Areas (Preservation and Enhancement) Act (1972), Public Parks Ordinance and the Botanic Gardens Ordinance.

New and amended legislation should provide for: structured integrated environmental planning and control; statutorily determined time-frames for completion and approval of national and sectoral plans; and adequate financial and legal incentives for sound agroforestry development in the country. This latter element is required to facilitate agricultural credit and loan facilities to encourage farmers. The consultant also advocates umbrella environmental legislation as is found in the Bahamas, Jamaica and St. Kitts and Nevis, and which is actively being considered in Guyana and Trinidad and Tobago. Such legislation should empower the Historical, Conservation and Environmental Commission as a statutory body with the necessary functions in this regard.

Many of the statutes cover the same areas, while many issues are not covered by any existing legislation. As in other CARICOM countries, this situation has arisen because legal development has tended to deal with single issues, due to perceived necessities or pressing concerns. This patchwork of fragmented and uncoordinated legislation is not rooted in any central theme and has contributed to confusion among agencies, non-enforcement of regulations and a dilution of responsibility. It is hoped that the GOAB will be committed to halting environmental degradation and to improving the quality of life for its citizens as well as the quality of enjoyment for its visitors. It should therefore move urgently towards rationalization of environmental legislation to ensure adequate provisions for protection and development of forests, natural habitats and wildlife in the interest of conservation of the environment and biological diversity.

Administration

Forestry falls under the portfolio of the Ministry of Agriculture, Fisheries, Lands and Housing and is staffed by a professional officer, two sub-professionals, four forest rangers, a nursery technician and a trainee. This is notable growth for the country which has never had the benefit of such capacity, even in colonial days.

The GOAB now needs to structure this team into a proper Forestry Division by providing it with clear functions and responsibilities and with adequate facilities to perform these duties effectively. This will require the promulgation of an official forest policy which, until now, has been notably lacking.

Constraints Affecting the Subsector's Development

As emphasized in many reports during the past decade, the main constraint is the lack of a coherent land use policy to guide the allocation of land for suitable uses and to provide security of tenure for those in dire need of adequate land space. Consequently, there has been no survey nor gazetting of forest reserves for fifty years since the Forestry Act and hence, there is no forest estate in Antigua and Barbuda. The absence of a land use policy has prevented the development of afforestation schemes and will be a serious hindrance to promoting agroforestry in the country, if not addressed urgently. At the same time, there is no forest policy nor the financial means to support one. However, the Government has moved steadily to provide trained personnel, although it has taken many years.

A potential constraint to the implementation of the NFAP could be the profusion of overlapping jurisdictions caused by the existing enactments. However, the dialogue initiated by the Historical, Conservation and Environmental Commission and the apparent coordinated support engendered by the TFAP suggest that suitable rationalization of these overlaps can be achieved.

OVERVIEW OF PRIORITY AREAS

Forestry in Land Use

Land Use Distribution: Present land use distribution is difficult to quantify reliably, as various estimates differ. The Land Use/ Watershed Management consultant has summarized previous percentage estimates as shown in Table below.

Table : Antigua & Barbuda Land Use Distribution.

CATEGORY	ANTIGUA LAND AREA (%)	BARBUDA LAND AREA (%)
Woodland	34	66
Mangrove Swamp	3	22
Scrub/Rough Grazing	33	9
Improved Pasture	4	-
Agriculture Crops	8	1
Settlements	11	-
Tourism	3	2
Airfields	3	-
Reservoirs	1	-

Rural land use activities include beekeeping, livestock, small cropping, large specialty farms and part-time farms. The main crops are pineapples, cotton, onions, pawpaw, coconuts and flowers. There is considerable opportunity to increase local production of many tropical fruits and to reduce imports and outflows of foreign exchange. This strategy can be integrated with a livestock improvement programme to reduce imports of meats. The feed for livestock would be enhanced by fodder and crop residues, while the animal litter would enrich the soils and improve their productivity. Numerous reports on land use have pointed to the need for watershed rehabilitation to enhance the ground water storage, reduce wind desiccation, increase fodder and fencing material. NGOs have become active in advocating tree planting for food production and amenity of the countryside, employing women and community groups as well as school children. In this regard, the Pan American Development Foundation and the Environmental Awareness Group have embarked on a small scale agroforestry project, working with about twelve communities to establish fruit orchards, live fencing and woodlots.

In Antigua the Government owns sixty percent of the land, some of which is currently rented to small farmers on a year-to-year basis. This system is practically feudalistic and offers no stability to the farmers who need to feel that they have a stake in their nation. It should be replaced with longer term leases and also by a system of freehold through a land settlement scheme based on economically sized parcels. Indeed, GOAB should consider divesting its farms and stations to the highest bidder, or to staff and workers in part settlement for voluntary redundancy. This would relieve pressure on the Recurrent Budget, while allowing inflows of cash to support necessary investments or loan repayments.

Watershed Management

Antigua and Barbuda depends mainly on surface runoff for its water needs: domestic, industrial, irrigation and recreational. Existing ground water resources are extremely limited, and reportedly are threatened by excessive use, as well as potential contamination from salt intrusion, chemical pollutants and inadequate sewage and waste disposal facilities. While efforts are in train to cope with this problems, attention needs to be given to measures that will reduce siltation of reservoirs and maintain the land in a conducive condition for optimal percolation and soil water retention.

There is very little data on surface runoff and infiltration, but over the years, the change in vegetation has obviously affected the volume of runoff and the transport of sediment into the reservoirs and ponds. Additional water sources include wells, the desalinization of sea water and cisterns. Total demand is 3-4 mgd (million gallons per day) of which actual consumption was 2.42 mgd in 1987. Delivery capacity is 2.2-3.8 mgd, there being considerable loss from broken pipes. Future demand is estimated at 5.1 mgd in the year 2000 and approximately 8.0 mgd in 2020. A proper water plan is needed particularly in view of the concentration on tourism and the increasing visits of cruise ships which combined, can take away from Antigua approximately 500,000 to 750,000 gallons of water per month.

While interest in watershed areas is primarily focused on water quality and quantity for storage sites, this may shift with increasing demand for food, to include crop production along with systems to minimize erosion and pollution. Roaming livestock have compacted the hillside lands and have browsed on cultivations and gardens.

As a possible solution for such a problem, it is recommended that agroforestry (multi-cropping) systems be used (eg. plant leucaena in contour hedges with cultivated grasses). Also, the regular illicit burning of rough grazing and woodland causes the spread of citronella grass and creates a greater problem for land rehabilitation. Before embarking on any watershed programme, a detailed plan of the area should be developed, including the location of agroforestry treatments, conservation measures, subdivision of farm holdings, protection of runoff and drainage channels.

Agroforestry

Within the framework of integrated land use planning and watershed management, the production of agricultural crops in combination with trees is an excellent conservation practice which offers the opportunity for sustained use of land to maximize returns, while protecting soil and water resources. In Antigua, agroforestry is not new, having been practised historically on large and small holdings. However, there is room for intelligent modifications to be applied according to local situations, and the approach is being proposed as a means of increasing land utilization and farm output, particularly fruits and livestock, primarily for the local market. In addition, meaningful proposals have been made for solving the problem of uncontrolled grazing, while providing forage, fencing, shelter and pasturage for livestock.

Forest Management and Silviculture

The present forest cover of Antigua is secondary forest, which has developed following the decline of the sugar industry. The most important species are Spanish oak (Inga laurina), silk cotton (Ceiba pentandra), mahogany, turpentine (Bursera simaruba), white cedar, and logwood (Haematoxylon campechianum). The mangrove forests consist of Rhizophora mangle. Avicennia nitida and Laguncularia racemosa. The dry scrub forests are dominated by Acacia spp. and Prosopis chilensis which were introduced and have become naturalized. Other introduced species include casuarina, neem (Azadirachta indica), tamarind (Tamarindus indica), leucaena (Leucaena leucocephala) and woman's tongue (Albizzia lebbeck).

It is not expected that the secondary forest will contain many trees of marketable size, and indeed there is no sawmilling capacity on the island. However, on the few plots of planted trees (neem, casuarina, leucaena, mahogany and white cedar), the growth has been promising, and annual measurement of selected stems on each site can be made by the Division's staff on a systematic basis. The data derived will assist in future determinations of species suitability and potential performance, should a large reforestation programme be adopted.

The Forestry Division can now be assigned responsibility for the management of Antigua's forest estate (when demarcated and gazetted) and for implementing a forestry programme based on the recommendations of previous studies and the NFAP. Existing forest cover and scrubland should be assessed in terms of the opportunity to conserve biodiversity or, where possible because of soil depth, to be reforested for the production of commercial timber and other products. This will entail air photo interpretation ground-truthing as a normal Divisional function, and preparation of maps and working plans. Forest rangers should be deployed in the field on regular patrols against illicit extraction, cattle trespass, clearing, fires, etc.

Forest Based Industrial Development

There is no industrial forestry activity apart from small scale charcoal production (largely for tourist and domestic recreation pursuits), extraction of wattle for fish and lobster traps, and some use of mahogany and white cedar for boat construction and fence posts. Estimates are that 1135 tonnes/year of charcoal and 4000 fence posts/year are produced, but there are no supporting data. The wood related manufacturing sector is also small, but again there are no records. Some opportunity may exist for handicraft material for carving, but technical training would be required to enhance growth in this area, particularly to utilize the limbs pruned from urban mahogany trees, some of which are dying or so badly mutilated that they should be felled and replaced. In addition, there is potential for non-wood products of agroforestry such as fruits and nuts, fodder and water.

Neem has been quite successful in Antigua and Barbuda and is widespread, mainly along road-sides or fence lines. It seems to have withstood the effects of hurricane Hugo in 1989, and attention should be given to determining the utilization prospects for this valuable fast-growing species. It is being planted elsewhere in the world and produces a wide range of products. These include the extract from the seed kernel and seed coat to yield neem oil, which is used as a lubricant, insect repellant and insecticide and in the manufacture of soaps and cosmetics. The seedcake residue is a valuable organic manure and cattle feed, while the root exudate has medicinal properties; the bark yields an industrial gum and the leaves are used as a green manure. The timber is also useful for lumber and fuelwood. Antigua and Barbuda has an excellent opportunity to investigate the industrial potential of this tree which could attract investment to the country.

Fuelwood and Energy

In spite of the relatively high per capita income levels in Antigua and Barbuda, there is still some consumption of charcoal. Unfortunately, no reliable records exist to indicate the levels of production and consumption. Charcoal burning does not presently constitute a threat to forest resources as suitable material is already available from pastures in the form of Acacia. This tree is a pest which threatens the health of cattle, as the thorns cause severe skin lesions, exposing the animals to the risk of disease. The GOAB is obviously aware of the continuing world energy crisis, and a case could therefore be made for encouraging the use of alternate sources of energy to save foreign exchange and to utilize natural products generated in the course of development initiatives such as agroforestry and watershed rehabilitation.

Conservation of Forest Ecosystems

In spite of its small size, the island possesses considerable wildlife and ecosystem diversity, much of which could enhance the growth of tourism but, more importantly, could awaken society's consciousness and pride in its natural environment. The Country Environmental Profile has recommended specific elements to be included in a system of national parks and protected areas, and these should be expanded to include coastal zones and forested areas.

Negative development pressure on the coastal zone can compromise the future ability of the country to sustain long-term socio-economic growth. All development proposals should be carefully reassessed to avoid adverse environmental impact. In the meantime, interim action can be taken to stimulate financing of resource management through collection of fees by local guides and monitoring of visitor levels and impact. Resource inventories are proposed under the NFAP projects.

Institutions

The main institutions concerned with forestry and environmental management include the Forestry and Fisheries Divisions, the Development Control Authority (DCA), the National Parks Authority (NPA) and NGOs, primarily the Environmental Awareness Group (EAG). Under the present tight fiscal policies, the public agencies are restricted in their ability to execute programmes. There are no funds available to undertake work programmes in forestry or agroforestry, to maintain monitoring of fisheries and fishermen, nor to expand the capability of the

NPA. The NGOs led by the EAG, are particularly knowledgeable and committed, and though lacking in resources, they are quite resourceful. They have played a significant role in raising environmental consciousness, although their confrontational approach has been met with official suspicion. However, they are now widely accepted in official circles that recognize the desirability of closer collaboration. This has led to the appointment of the Historical, Conservation and Environmental Commission (HCEC), aimed at improved dialogue and coordination within the public sector. However, the general institutional situation of the public sector is weak in implementing capacity and performance, and the NGOs are still skeptical that the Government will change its ways.

The NPA is backed by strong legislation which provides for inter-sectoral linkages to consult and cooperate with agencies having similar aims and objectives. It also has the power to delegate its functions and this could be done through the Barbuda Council, Forestry and Fisheries Divisions, the EAG and other NGOs.

The Barbuda Council was established by the Barbuda Local Government Act (No. 15, 1976), to manage Barbuda's government affairs. The Council consists of 9 members who elect a chairman and secretary from among them. Proposals for capital and special expenditure are required to be submitted to the Ministry of Home Affairs (Central Government) who forward them to the Ministry of Finance. The Council has powers to make bye-laws for development on the Island.

The conservation and development of forest resources can only be implemented by capable resource management institutions, properly staffed and financed, and provided with the necessary authority for efficient execution of their responsibilities. Unfortunately, the lead agencies in Antigua and Barbuda do not seem to be adequately equipped at present, and will require considerable strengthening to improve their management capacity.

With the recent return of the country's first professional forester, the Government now possesses the nucleus of trained forestry personnel. The Forestry Division can now be empowered with the following responsibilities and functions:

(a) protection of the forest estate, wildlife sanctuaries and protected areas;

(b) afforestation and forest management in watershed areas;

(c) cooperation with private and Government agencies on urban forestry programmes (including the Botanical Gardens), forestry education and information;

(d) assessment and conservation of biodiversity and promotion of ecotourism;

(e) generation of self sustaining activities utilizing forest and non-wood products;

(f) preparation of budget plans and annual performance reports.

The following steps are therefore recommended for official approval and action:

(i) adoption of a formal national forest policy and appointment of the Chief Forestry and Wildlife Officer;

(ii) declaration and gazetting of the forest estate, followed by survey and demarcation of forest reserve boundaries;

(iii) preparation and approval of a baseline work plan for the Division, including job descriptions for all staff;

(iv) transfer of game wardens, caretakers and necessary staff to the budget of the Division;

(v) assignment of responsibility for supervising the maintenance of the Victoria Park and Botanical Gardens;

(vi) assignment of office accommodation, vehicles and equipment to enable efficient performance of the Divisions functions.

After establishment of the Division, there are a number of basic activities which it can undertake, namely:

- planning, layout and implementation of a forest recreation centre at Wallings;
- planning and implementation of urban forestry based on the Botanical Gardens;

- field assessment of planted forested trees, enumeration, diameter and height measurements, site conditions;
- assessment of charcoal production from leucaena at Dunbars;
- continued cooperation with Fisheries Division and NGOs in mangrove and wildlife monitoring;
- initiation of contacts with NPA to discuss areas of collaboration;
- selection of possible reforestation sites and planning proposals for employment absorption.

These activities can be undertaken with little or no new financing and are within the technical capability of the trained staff. The results will soon be evident if tasks are approached in a business-like manner and with an eye to the future.

PLAN OF ACTION

Based on the results of the Country Mission, as outlined in the previous sections of this report, and on the working documents prepared by various consultants, a Plan of Action is proposed for the development of the country's forestry subsector. The National Forestry Action Plan (NFAP) proposes elements of national development policies, a long-term policy for the sub-sector, together with priority area policies and objectives on which specific action projects are proposed for a five-year period.

Mindful of the country's present financial situation, the Plan proposes redeployment of the existing personnel and budget as the Government's initial investment in the sub-sector to attract donor support in setting a sound foundation for sustainability. The plan includes the normal work of the Forestry Division which will also function as lead agency in the implementation of three projects.

The NFAP is estimated to cost a total of US $3.0 million of which US $2.2 million would be contributed by international donors while the remaining US $0.8 million would be contributed by Antigua and Barbuda, mainly in kind.

National Development Policies

The adoption of national development policies for land use planning, land allocation and sustainable resources management are recommended for enhancing the contribution which forestry and trees can make to socio-economic development and environmental stability.

Long-term Subsector Policy and Objectives

59. It is proposed that the long-term Forest Policy of Antigua and Barbuda should be:

(a) to protect the country's natural forest resources;

(b) to promote the development of forests and tree cover on suitable lands;

(c) to encourage the application of agroforestry in planned land use for rural development;

(d) to pursue opportunities for utilization of forest and other products.

In pursuing this forest policy, the Government will seek to ensure:

- the wise use and proper management of the island's limited forest and natural resources;
- the preservation of representative samples of the island's natural forest and vegetation resources and the maintenance of the ecological balance;
- the provision of opportunities for the public for scientific, educational and recreational activities to enhance knowledge of forest ecosystems, research in potential utilization of the gene pool and tourism benefits;
- the stimulation of industries capable of boosting the island's economy, based on products derived from forests and trees.

Therefore, the proposed objective of the NFAP is to establish and manage a national forest estate on Government and private lands. This will ensure conservation of soil and water resources, protection of forest ecosystems and production of wood and related products for social and economic benefits on a sustainable basis.

Priority Area Policies and Projects

In the medium term, the NFAP will be guided by the priority area policies and strategies outlined below in the implementation of the respective action projects. These policies are all inter-related and represent the minimum actions required to effect the fundamental changes necessary towards achieving the basic aims of the new long-term forest policy.

Institutions-to provide effective environmental legislation and capable institutions for proper administration and sustained development of forestry and related programmes, with the participation of local people and interest groups.

The purpose of this policy is to improve the environmental management-capacity of the country and to install multi-sectoral and participatory mechanisms for planning and monitoring of forestry and related programs. This aspect is of the highest priority if the national long-term policy and objective is to be attained. Accordingly, the following project is proposed for immediate attention:

Country Capacity for NFAP Implementation

The objective of the project is to build capacity in the lead agencies to ensure initiation and co-ordination of NFAP implementation, sustainable natural resources administration and management, project preparation and management, public education and extension, and environmental management for conservation of natural resources.

The project includes specific provision for the survey and demarcation of forest reserve boundaries in order to create a forest estate.

Forestry in Land Use-the proposed policy recognizes the multiple benefits which accrue from existing tree cover and seeks to promote forestry and agroforestry as appropriate land uses for scrub lands and areas in need of conservation.

In pursuing this policy, the aim will be to allocate an adequate area of suitable land to forestry and agroforestry to ensure conservation of soil and water resources, sustainable income and employment for rural people, and integrated crops and livestock production. This will require a programme of land use survey, classification and planning to effect proper subdivision and leasehold settlements as incentives towards encouraging greater tree coverage on farms.

Forest and Watershed Rehabilitation

The objective of this project is to establish tree cover on degraded watersheds and sand-mined lands in order to reduce erosion, improve water retention and enhance rural development through agroforestry practices and reforestation.

Forest based industrial development-the proposed policy seeks to ensure production of future supplies of wood raw material and utilization of existing mature timber for industrial purposes, while maintaining the amenity value of trees in urban areas.

Utilization of Neem

The objective of this project is to study the prospects for utilizing various products of the neem tree with a view to industrial development.

Fuelwood and energy-the proposed policy seeks t-explore opportunities to utilize fuelwood for energy purposes and to promote feasible wood energy applications in appropriate industries.

The aim would be to develop a sustained demand for wood energy by domestic and industrial users, whether as fuelwood or as charcoal, both to promote the use of agroforestry products and to conserve other energy sources.

The objective of this project is to determine the production and demand for charcoal and to evaluate the opportunity to expand the market.

Conservation of forest ecosystems-this policy seeks to maintain biological diversity and to protect forest ecosystems and natural habitats, while sustaining the flow of social, economic and environmental benefits derived from forests for nationals and tourists.

The aim is to increase public awareness of environmental problems and the need to conserve the natural resources of the island as part of the nation's heritage. In the absence of a coordinated system of management of existing habitats, the following project is extremely important:

Establishment of National Parks/Protected Areas System

The long-term objective of the project would be to ensure conservation and management of natural ecosystems and protected areas, to maintain biodiversity and to stimulate research in natural products and public appreciation for the environment.

Plan Coordination and Implementation

The Antigua and Barbuda NFAP consists of three projects to be implemented by the Ministry of Agriculture, one of which will be in cooperation with the NPA and the EAG. In view of the need for inter-sectoral cooperation and public participation, a suitable coordinating mechanism will be necessary.

It is therefore recommended that the TFAP National Coordinating Committee be replaced by a NFAP Steering Committee, under the chairmanship of the Permanent Secretary, with somewhat reduced membership.

Coordination of the NFAP will be needed at both the national and the operational levels.

(a) At the national level, the steering committee will be responsible to the Minister for directing project preparation, examining project documents and environmental assessments and setting guidelines for timely project implementation. The Minister will steer approvals through the Cabinet where necessary.

(b) At the operational level, the steering committee will require special secretariat and follow-up services. Therefore a

- NFAP Projects Unit should be established in the Ministry, headed by a Project Co-ordinator and supported by an Economist/Planner and a Secretary. As the focal point of contact, the Project Coordinator would be responsible to the Chairman of the Steering Committee for preparing the committee's agenda, while coordinating all inputs necessary at each stage of NFAP implementation.

The development of institutional capacity for implementation of the projects could benefit enormously from technical assistance, by way of a Chief Technical Advisor, to assist in harmonizing the complexities and priorities of environmental management.

The GOAB should examine its needs urgently and initiate early contact with potential international and bilateral donors regarding the possibility of supporting a Country Capacity Project (CCP).

The personnel proposed for each project has been kept to a minimum by identifying the skills currently available in the country, anticipating that suitable personnel will be assigned from their respective agencies and divisions to assist in project implementation.

Where expertise has to be recruited from abroad, every effort should be made to obtain CARICOM experts, and also to secure TCDC (technical cooperation among developing countries) wherever possible.

Table : Foreign Personnel Requirements for NFAP

FOREIGN PERSONNEL REQUIRED	PERSONNEL NUMBER	TIME NEEDED (MM [1])	PROJECT NUMBER
Project Manager	1	60	2
	1	36	3
Soil Conservation Spec.	1	36	2
Utilization Specialist	1	12	2
	1	6	2
	1	6	1
Parks Planner	1	6	3
Chief Technical Adviser	1	36	1
Legislation Specialist	1	6	1

1 MM = Man Months

Table : Local Personnel Requirements for NFAP

LOCAL PERSONNEL REQUIRED	PERSONNEL - NUMBER	TIME NEEDED (MM)	PROJECT NUMBER
Forestry Assistant	1	60	2
	1	36	3
Livestock Officer	1	60	2
Extension Officer	1	60	2
Economist/Planner	1	36	1
Terrestrial Biologist	1	24	3
Fisheries Officer	1	36	3
Project Coordinator	1	60	1
Land Surveyor	1	24	1
Foreman/Field Assist.	1	60	2
	1	6	2
	3	108	3
Secretary	1	60	2
	1	36	3
	1	60	1

IMMEDIATE FOLLOW-UP

In preparation for the upcoming round-table meetings, the TFAP National Coordinating Committee should immediately review the draft project profiles and obtain the Government's endorsement of the Plan.

At the same time, with the support of the Permanent Secretary, Ministry of Agriculture, the TFAP National Coordinator should make informal contact as soon as possible with potential donors regarding their likely interest in providing early support for the proposed CCP. It is essential to maintain the momentum and interest that was generated during the country mission, so that the other projects can be further fine-tuned before the national round-table.

Although the final shape of the NFAP will be based on the outcome of the round-table meetings, its refinement for implementation will depend initially on the results of the land use survey, and on the approval by the Minister (or an inter-ministerial council) of the Steering Committee's subsequent recommendations.

DRAFT PROJECT PROFILES

Summary of NFAP Budget by Priority Areas

Table : Summary of NFAP Budget by Priority Areas (US$ X 1000)

PRIORITY AREA	NO.	DONOR US$$ (X 1000)	GOAB [1] US$$ (x 1000)	TOTAL US$$ (x 1000)
Institutions	1	529	150	679
Land Use	2.1	892	368	1,260
Industries	2.2	110	4	114
Fuelwood	2.3	100	5	105
Conservation	3.0	455	205	655
TOTALS		2,086	732	2,813
% of TOTAL		74%	26%	100%

1 Government of Antigua & Barbuda

Proposed Implementation Schedule

Table : Proposed Implementation Schedule.

PROJECT NO.	(S = start)		TIME in YEARS		(E = end)	
	1	2	3	4	5	6	7
1.0	S					E	
2.1		S					E
2.2		S	E				
2.3			S	E			
3.0		S			E		

COUNTRY CAPACITY FOR NFAP IMPLEMENTATION

PRIORITY AREA	Institutions
DURATION	5 years (due to shortage of forestry staff in terms of numbers and experience)
LOCATION	Country wide
TARGET BENEFICIARIES	Ministry of Agriculture, National Parks Authority and NGOs.

Immediate Objectives

To implement the steps necessary for successful initiation of NFAP projects, and to establish appropriate institutional capacity and mechanisms for effective project implementation and management.

Activities

1. Preparation of implementation plan and project documents in keeping with donor requirements.
2. Coordination of project inputs.
3. Amendment, integration and consolidation of related enactments and mechanisms.
4. Support for lead implementing agencies.
5. Boundary surveys of forest reserves and protected areas.
6. Public education and information, including herbaria, research.

Outputs

1. Technical assistance and funding of projects.
2. Implementation of planned projects.
3. National policies and institutional mechanisms for improved environmental management and development.
4. Increased NGO and public participation.
5. Establishment of the national forest estate and protected areas.

Inputs

Equipment and materials	Personnel
Microcomputer w. GIS	Chief Forestry Adviser
1 x 4wd station wagon	Project Coordinator
2 x 4wd pick-ups	Economist/Planner
2 x portable radios	Land Surveyor
	Secretary
	Legislation Consultant
	Utilization Consultant

Training

Professional Level:

-2 Professionals

Sub-Professionals:

-4 with Certificates

-4 with Diplomas

National Lead Implementing Agency

Ministry of Agriculture, Fisheries, Lands and Housing.

Linkages with Other Projects

NFAP 2.0, NFAP 3.0

Observations

This project is of the highest priority for the NFAP, and should be initiated as soon as possible.

Project Budget

Table : Personnel with Cost Requirements[2]

PERSONNEL	MAN MONTHS	FOREIGN COSTS (US $$)	LOCAL COSTS (EC $$)
Chief Tech. Assistant	36	330,000	
Project Coordinator	60		150,000
Economist/Planner	36		75,000
Land Surveyor	24		50,000
Secretary	60		80,000
Legal Consultant	6	45,000	
Utilization Consultant	6	45,000	
SUB-TOTAL	228	420,000	355,000

2 Project I with priority I rating.

Table : Equipment with Cost Requirements

EQUIPMENT	FOREIGN COSTS (US $$)
Micro-computer & GIS	25,000
Vehicles (3)	70,000
Portable Radios (4)	4,000
SUB-TOTAL	99,000

Table : Practical Aspects with Associated Costs

PRACTICAL ASPECTS	FOREIGN COSTS (US $$)	LOCAL COSTS (EC $$)
Training	10,000	
Office & Field Operation		50,000
SUB-TOTAL	10,000	50,000

Table : Total Budget for Project

BUDGET ITEM	FOREIGN COSTS (US $$)	LOCAL COSTS (EC $$)
PERSONNEL	420,000	355,000
EQUIPMENT	99,000	
PRACTICAL ASPECTS	10.000	50,000
TOTAL BUDGET	529,000	405,000

FOREST AND WATERSHED REHABILITATION WITH A SUSTAINED UTILIZATION COMPONENT

PRIORITY AREAS	Forestry in Land Use/Industries/Fuelwood
DURATION	5.0 years
LOCATION	Body Ponds, Dunnings, Christian Valley, Bethesda, Monks Hill, John Hughes Liberta/ Swetes/Wallings, Falmouth/English Harbour, Dunbars, Pot Works Dam perimeter, Palmetto Point (Barbuda).
TARGET BENEFICIARIES	Watershed Inhabitants, Land Owners, Domestic Water Consumers, Charcoal Producers, Tourists, Farmers.

Immediate Objectives

1. To establish tree cover on degraded watershed and sand-mined lands in order to reduce erosion, improve water retention and enhance rural development, through agroforestry conservation practices.
2. To study the potential for utilizing neem to produce insecticides and fertilizers for local and foreign markets.

3. To determine the current production capability for charcoal and to investigate the possibility for increasing output and demand.

Activities

1. Prepare comprehensive Land Use Plan based on the physical assessment of available sites.
2. Assess existing tree cover, orchards, livestock herds, and investigate potential yields and markets, and returns.
3. Institute fire protection, livestock integration/control and forest extraction control procedures.
4. Establish trial plots of selected tree species (neem, etc.).
5. Obtain financing and commence implementation of seedling production, reforestation, watershed management, agroforestry measures including a maintenance programme.
6. Organize field training for farmers, workers, and communities.
7. Update data base on rural sociology and prepare land allocation plan for farmers/settlers.
8. Evaluate work plan results and prepare reports.
9. Upgrade seed collection and nursery production with a maintenance component.
10. Review technology and applications of neem and prepare a feasibility study for its plantation, harvesting and marketing.
11. Conduct a survey of all charcoal burners and their productivity; demonstrate an improved kiln design.
12. Assess opportunities for linkages with tourism and environmental concerns.

National Lead Implementing Agency

Ministry of Agriculture, Forestry Unit in collaboration with the Barbuda Council, Antigua Public Utilities Authority, NGO's.

Observations

High priority project with the opportunity for use of reassigned casual employees from various Ministries.

Outputs

1. Optimal Land Utilization Plan covering Government properties, private lands, infrastructure and services.
2. Promotion and encouragement of agroforestry opportunities.
3. 250 ha of reforested lands.
4. 500 farmers trained in agroforestry techniques.
5. Control of bush fires and roaming livestock.
6. Update data base for rural development.
7. Data on species and programme performance.
8. Protected soil and water resources and improved water and agricultural yields.
9. Assessment of the industrial potential to produce neem tree products.
10. Assessment of fuel wood feasibility with the alternative energy potential.

Inputs

Equipment and Materials	
4 w.d. Station wagon (1)	4 w.d. Pick-up (1)
Survey instruments	Aerial photography, maps
Seeds, fencing	Tools, radios
Planting materials	Chain saw, axes, wedges
Kiln and scale	

Personnel	
Project Manager/Forestry Expert	Forestry Assistant
Soil conservation specialist	Livestock Officer
Extension Officer	Secretary
Foreman/Field Assistant (2)	Wood energy expert
Neem utilization expert	Field Assistant

Project Budget

Table : Personnel with Cost Requirements 3

PERSONNEL	MAN MONTHS	FOREIGN COSTS (US $$)	LOCAL COSTS (EC $$)
Project Manager	60	500,000.00	
Soil Conservation Specialist	36	270,000.00	
Forestry Assistant	60		90,000.00
Livestock Officer	60		125,000.00
Extension Officer	60		125,000.00
Foreman/Field Assistants	66		82,000.00
Secretary	60		80,000.00
Wood Utilization Expert	12	100,000.00	
Neem Utilization Expert	6	45,000.00	
SUB-TOTAL		915.000.00	502.000.00

3 Project 2 with priority 2 rating.

Table : Equipment with Cost Requirements

EQUIPMENT	FOREIGN COSTS(US $$)
4 w.d. vehicles (2)	50,000.00
Tools/instruments, saw	60,000.00
Radios (2)	2,000.00
Planting material	25,000.00
Fencing	25,000.30
Field trials	10,000.00
SUB-TOTAL	172,000.00

Table : Practical Aspects with Associated Costs

PRACTICAL ASPECTS	FOREIGN COSTS (US $$)	LOCAL COSTS (EC $$)
Operating expenses		16,000.00
Field operations		500,000.00
Training	15,000.00	
SUB-TOTAL	15,000.00	516,000.00

Table : Total Budget Requirement

BUDGET ITEM	FOREIGN COSTS (US $$)	LOCAL COSTS (EC S$)
EQUIPMENT	172,000.00	
PRACTICAL ASPECTS	15,000.00	516,000.00
PERSONNEL	915,000.00	502,000.00
TOTAL BUDGET	1;102,000.00	1,018,000.00

ESTABLISHMENT OF A NATIONAL PARKS/PROTECTED AREAS SYSTEM

PRIORITY AREA	Conservation of Forest Ecosystems
DURATION	3 years
LOCATION	Wallings, Dark Wood, Gaynors, Great Bird Island, Goat Island (Barbuda), Bird Sanctuary (Codrington Lagoon, Barbuda).
TARGET BENEFICIARIES	Nationals and Tourists

Immediate Objectives

To plan and establish a system of nature reserves, sanctuaries and protected areas including mangroves for conservation of biodiversity, public education regarding the natural environment and ecotourism development.

Activities

1. Gazetting of proposed sites and buffer zones.
2. Field surveys and data collection.
3. Classification and determination of use specifications.
4. Preparation of management plans.
5. Training of technical field staff and local persons.
6. Development of interpretive and educational programmes.
7. Planning and layout of visitor facilities, trails.

Outputs

1. Antigua and Barbuda national parks system.
2. Improved environmental management and monitoring capability.
3. Increased number of visitors and national environmental awareness.
4. Increased revenues and linkages with tourism.

5. Data bank of natural resources.
6. Guidelines for resource management and development.

Inputs

Equipment and materials	Personnel
1 x 4wd station wagon	Project Manager
1 x 4wd pick-up	Parks Planner
1 x 15 foot boat w. motor	Terrestrial Biologist
1 x rubber dinghy w. motor	Marine Biologist
Sampling aids, etc	Fisheries Officer
2 x portable radios	Forestry Assistant
	Secretary
	Game wardens

National Lead Implementing Agency

National Parks Authority in cooperation with Forestry and Fisheries Division, and EAG.

Linkages with Other Projects

NFAP 1.0 (Country Capacity Project)

Observations

This project should be coordinated with similar projects in the Caribbean and Central American region.

Project Budget

Table : Personnel with Cost Requirements

PERSONNEL	MAN MONTHS	FOREIGN COSTS (US $$)	LOCAL COSTS (EC $$)
Project Manager	36	300,000	
Parks Planner	6	45,000	
Terrestrial Biologist	24		50,000
Fisheries Officer	36		75,000
Forestry Assistant	36		40,000
Game Wardens (3)	108		160,000
Secretary	36		48,000
SUB-TOTAL	282	345,000	373,000

Table : Equipment with Cost Requirements

EQUIPMENT	FOREIGN COSTS (US $$)
4 wd Vehicles (3)	70,000
Boats & Engines (2)	23,000
Portable Radios (2)	2,000
Sampling Gear	5,000
SUB-TOTAL	100,000

Table : Practical Aspects with Associated Costs

PRACTICAL ASPECTS	FOREIGN COSTS (US $$)	LOCAL COSTS (EC $$)
Training	10,000	
Operating Expenses		180,000
SUB-TOTAL	10,000	180,000

Table : Total Budget

BUDGET ITEM	FOREIGN COSTS (US $$)	LOCAL COSTS (EC $$)
PERSONNEL	345,000	373,000
EQUIPMENT	100,000	
PRACTICAL ASPECTS	10,000	180,000
TOTAL BUDGET	455,000	553,000

4

Capacity Building in a Forest Project, India

SCOPE

The purpose of the biodiversity case study is to draw and present lessons learnt from the preparatory phase of the project "Capacity Building for Participatory Management of Degraded Forests in Orissa, India" regarding the integration of biodiversity aspects in the project. The study attempts to analyse to what extent biodiversity has been integrated into the project, which biodiversity aspects that have been considered, the mechanisms for integrating biodiversity aspects, and the relevance of these to the different stakeholders.

It is hoped that the lessons learnt will give some guidance regarding:

- methods for how aspects of biodiversity can be integrated into both the assessment phase of a project/programme and on-going projects/ programmes, and
- how to monitor aspects of biodiversity in the programmes/ projects.

METHODS

The study has been undertaken through:

- studying reports and Sida documents regarding lessons learnt from the Sida-supported Social Forestry Programme in Orissa;

- studying the different reports from phase one of the project
- consultation with programme officers concerned at Sida-DNRE (Anita Ingevall and Ramesh Mukalla);
- drawing lessons from previous consultations with project staff (Dr L A K Singh, Ulf Öhman, Peder Nilsson, Dr R. V. Singh, Ajay Rai and others);
- drawing lessons from previous consultations with men and women from 12 local communities in Orissa in Mayurbhanj, Dhenkanal and Cuttack districts in Orissa (as part of the biodiversity study for the project preparation);
- drawing lessons from previous consultations with representatives from NGO's (primarily Oxfam, Vasundhara, MASS, and PIPAR) and from a district level forest federation in Orissa (Mr Udanayath Mohanty, president);
- studying other background material of relevance for the case study.

The study was carried out during autumn 1999 and spring 2000 by Marie Byström. A draft was presented and discussed during a seminar at the Natural Resources Department at Sida in January, 2000.

Valuable comments were received from Dr L A K Singh, Orissa Forest Department, Anita Ingevall, Sida and Kees Manintvedt, ETC Ecoculture, the Netherlands. The terms of reference for the case study are found in Appendix 1. Based on the comments received, a final report was pre-sented in May, 2003. The developments in Orissa between 2000 and 2003 have not been analysed in this report.

BRIEF DESCRIPTION OF THE PROJECT AREA

Orissa is situated on the East coast of India. With 32 million people and 155 000 sq.km it is one of the least densely populated states of India. Thirty-eight percent of Orissa's population belongs to Scheduled Castes or Scheduled Tribes, which are generally regarded as 'marginalised groups' in Indian society. This is the highest proportion in the country (Statistics India, 1998). Orissa is also one of the least urbanised states with 88% of the people living in 50 000 villages. According to Statistics India (1998), 95% of the population are Hindus, and 76% speak Oriya. Of Orissa's land area, 58 000 sq.km is officially designated as forest, or about 37% of the total land area. Roughly 50% of this area would in a natural state be constituted by dry deciduous forests dominated by sal (Shorea robusta), 30% would be moist deciduous forests and 20% covered by semi-evergreen forests, littoral forests and tidal swamps (OFD,1999). To understand the forest conditions and forest administration in Orissa today, it may be helpful to look back over the last two centuries.

In 1855, under the British colonial rule of India, the Imperial Forest Department was created. Among the first actions of this Forest Depart-ment was the demarcation of forests to create Reserved Forests (RF) and Protected Forests (PF) in order to meet industrial demands (e.g. ship-building and later building of railways) of the British Empire. Rights of the majority of people who depended on the forest resources were diminished (Pal, 2000).

After Independence, the Indian government nationalised the Reserved and Protected Forests and adopted the British system of forest management. Today, all land classified as forest belongs to the State, and there is in principle no private ownership of forest land. The utilisation of forests and forest produce is governed by an intricate legal framework. Nevertheless, major illegal felling takes place. Forest degradation has been serious during the last decades.

Today, of the 58 000 sq. km of forests in Orissa, about 10 000 sq. km forest land is estimated by Orissa Forest Department (OFD) to be devoid of any vegetation, and the area of degraded forests is estimated at a total of 31 000 sq.km (OFD, 1999).

However, forests and forest products are still very important for millions of people in Orissa. Of the total revenue earned from the forest sector in Orissa, around 80% comes from non-timber forest products (NTFPs) or Minor Forest Produce (MFP's) (Mallik et al., 1998). Here, it must be remembered that the current logging ban has greatly reduced revenues from timber products. The collection and trade of NTFP's has until 2000 been controlled by a few agencies which lease the rights to the produce from the State. Local people who collect NTFP's have had no rights to store quantities of the produce, sell it on an open market or process it in order to add value.

The degradation of Orissa's forests has led to protection of forests by local communities in all parts of the state. Today it is estimated that at least 4 000 sq.km of forest land is protected and managed by some 10 000 village communities (Mahapatra, 1999). This amounts to around 7% of the total forest area. The community based forest protection and management dates back to the 1930's in some villages. After tribal uprisings and demands for complete rights over forest resources in the 30's the kings in Mayurbhanj, Nayagarh and Bolangir shared their rights to forests with the communities, thus initiating what is today called joint forest management. Much of the forest protection in Orissa has been initiated and managed by communities without involvement of the State, and there is today a tension between what is termed community forest management (CFM) and joint forest management (JFM).

National Policy Development on Joint Forest Management

The concept of joint forest management (JFM) grew in India in the 1980's as a result of a growing realisation of the necessity to acknowledge people's right to foreste. JFM is today central in India's national forest policy. The National Forest Policy of 1988 provides a broad framework for JFM. Its main objectives are: (i) maintenance of environmental stability and restoration of ecologi-cal balance, (ii) conservation of biological diversity and genetic resources, (iii) checking soil erosion and denudation in catchment areas of rivers, lakes and reservoirs, (iv) checking the extension of sand dunes in desert areas and along the coastal tracts, (v) increasing sustainability of forest cover through massive afforestation and social forestry programmes, (vi) meeting timber and non-timber requirements of rural and tribal populations, (vii) increasing productivity of forests to meet essential national needs, (viii) encouraging efficient utilisation of forest produce, and (ix) creating a massive people's movement with the involvement of women for achieving the above objectives and to minimise pressure on existing forests (OFD, 1999).

The policy states that domestic requirements of village communities living near forests for fuelwood, fodder, minor forest produce and con-struction timber should be "the first charge on the forest produce". The Orissa forestry sector strategy is developed from the National Forest Policy. It charts an increased role for rural people in forest man-agement and protection and acknowledges the need to devolve increasing benefits to them in meeting their multiple requirements of forest re-sources. The first state level Government Resolution (GR) on JFM was published 1988. It was subsequently amended in 1990 and replaced in 1993. The 1993 GR elaborately articulated the roles and responsibilities of OFD and the village forest protection committees. It also listed mechanisms for development and approval of Joint Forest Management Plans, benefit sharing provisions (50% share to the protecting community of any major harvest), the right to collect the usufruct (non-timber forest products for domestic use) free of charge and the right to appropriate all intermediate yield from silvicultural operations.

The 1996 GR further elaborated these mechanisms and processes. However, the 1996 GR has in practice not been implemented. Its implementation will depend on strengthened capacity of OFD in developing mechanisms to this end, which will require major resources for regulation and monitoring. While OFD has facilitated the formation of village forest protection committees on a fairly large scale and initiated joint forest management in many areas, the benefit sharing provisions of the Government Resolu-tions have not been implemented to any significant degree. While village level committees we visited were aware of their 'usufruct' rights they were largely unaware of their rights to all intermediate yield from silvicultural operations and 50% share of major harvests. In addition, a general logging ban has made halted the production of timber from village protected forests (Scandiaconsult et al., 1998).

Some communities have little faith in OFD and claim full rights to the forests they protect. While some communities have established village forest protection committees with official JFM agreements with OFD, other communities prefer to protect their forests on their own, in com-munity forest management (CFM) arrangements (Conroy et al., 2000). There are many non-governmental organisations (NGOs) in Orissa that help initiate and support CFM and JFM. Communities and NGOs have also formed district level forest federations in four districts, and in 1999 a state level forum called Orissa Jungle Manch was formed by NGOs and district federations (Pal, 2000). Sida supports the community level forest protection movement in Orissa through direct support to local NGOs for networking and advocacy.

PEOPLE AND FOREST BIODIVERSITY IN ORISSA

In most communities in the forested parts of Orissa, forest products provide an important part of subsistence and income, particularly for the tribal groups. Trees provide wood for construction of houses, tools and fuelwwod. Non-timber forest products (NTFPs) from a large diversity of species are collected and used for domestic consumption or sold. Until 2000, the rights to trade and process NTFPs were leased in a monopoly contracting system to a small number of traders and compa-nies, some of them state owned. Local communities had no such rights and people were forced to sell their NTFPs to defined outlets, often for a pittance (Hobley and Shields, 2000). This situation has now changed, partly as a result of the PMDFO process (see below under section 6.3.). Both NTFPs sold and NTFPs used domestically are vital for the liveli-hoods of Orissa's forest dwelling communities (Mallik et al.,1998).

For example, green leaf vegetables, small game, fruits and tubers provide proteins, minerals and vitamins which would otherwise be insufficient in the local diets, and they are of key importance during food shortages. Mushrooms form an important part of the diet during the 2-4 months when they are available, particularly for tribal groups. Particularly tribal people may spend a few hours daily for mushroom collection during this period. In many of the villages, mushrooms are sold at the local market or to other households in the village (while tribal women may spend hours daily in the forest, non-tribal women in the same village may seldom or never enter the forest).

People regularly use medicinal herbs for treatment of cattle diseases and for minor ailments like skin injuries, colds, coughs and stomach disorders.

Villagers' views on Biodiversity Values and Functions

There is a vast body of knowledge on forest biodiversity among commu-nities in Orissa, i.e. knowledge of occurrence and distribution of plant and animal species and functions and uses of these (Singh et al.,1999). The communities visited in the FMDFO preparatory phase empha-sised the importance of ecosystem functions for forest health and for the health and fertility of surrounding areas (forests were seen as important for e.g. regulation of hydrological cycles and for provision of humus and nutrients for surrounding fields) (ibid.).

Both "use-values" and "non-use-values" of biodiversity were impor-tant in the communities visited. The communities were proud of their protected forest areas and cared for the forest not only for its production of goods but also (and explicitly in higher caste groups) for its existence value. Parts of some of the protected forests areas were sacred groves and carried important religious functions. Tribal people in the villages invari-ably said that life without the forest would be unthinkable. Their identity is linked to the forest.

A reference may be made here to all the different life forms in the forest, the animals, plants, fungi and micro-organisms. Just like the humans, the individual lives in the forest also have a stake in its contin-ued existence for their own being and regeneration. They may be re-garded as stakeholders in their own right.

PROJECT DESCRIPTION

In 1997, Orissa Forest Department embarked on preparation of the project titled "Capacity Building for Participatory

Management of Degraded Forests in Orissa" (in this report referred to as PMDFO) with funding from Sida. Scandiaconsult Natura AB and Asian Forest Network assisted Orissa Forest Department in the project preparation phase, which took place during the period January 1998-May 1999.

The PMDFO project aims at promoting sustainable and community based management and use of forests in Orissa and thus contribute to social and economic development in Orissa. The preparatory phase of the project (Dec 1997-May 1999) aimed at developing background and capacity for a longer support in the second phase of implementation. The goals of the preparatory phase as expressed in OFDs final inception report (Scandiaconsult Natura AB/Asia Forest Network, 1998a) were:

1. Further elaboration of JFM and CFM concepts by exploring relation-ships between joint and community based forest management prac-tices;
2. Strengthening the capacity, including restructuring, of the Orissa Forest Department;
3. Knowledge about low cost methods for reforestation through natural regeneration;
4. Use of thc information and learning experiences generated during this period for the preparation of the project document for phase II.

The expected results of the first project phase were:

(a) a project document and work plan for the next 3-5 year phase;

(b) reports from a number of studies related to the project goals;

(c) an established and operative policy cell within OFD;

(d) a plan for the reorganisation of OFD;

(e) a plan describing how to study the succession of natural regeneration and its management, and how to incorporate lessons learnt within OFD;

(f) training carried out for staff and village representatives.

The following studies were carried out in the preparatory phase of the PMDFP project:

- Qualitative assessment of community initiative in forest management
- Benefit sharing mechanism between OFD and communities
- Legal framework and institutional arrangement between OFD and communities
- Identification of research areas and preparation of prioritised research agenda and research approach based on participatory concept
- Study on actors of relevance for implementation of themes of the project
- Study on entry points of relevance for implementation of themes of the project
- Silviculture and management practices in community based forest management initiatives
- Study on biodiversity aspects
- Study on domestic and commercial use of NTFPs, excluding marketing
- Study on domestic and commercial use of NTFPs, including marketing
- Identification and testing of criteria for area selection in forests not protected by communities
- Monitoring and evaluation for CFM and JFM areas

Reports from the studies have been presented, and a base paper has summareised the resommendations from the studies (Scandiaconsult Natura AB/Asia Forest Network, 1998b).

A project document for the implementation phase has been presented to Sida for consideration of support (OFD, 1999). Due to delays of OFD reorganisation for various reasons during the preparatory phase, a policy cell within OFD has not been established but is proposed in the project document. The same applies to training of staff and village representa-tives. The implementation phase of the project "Capacity Building for Participatory

Management of Degraded Forests in Orissa" has not yet started, and funding is still sought. A complicating factor here is the withdrawal of funds for Swedish development cooperation with India after the Indian nuclear bomb tests in 1998.

ANALYSIS OF LESSONS LEARNT IN RELATION TO BIODIVERSITY ASPECTS OF PMDFO

As in any complex project development process with many actors in-volved, the lessons that can be drawn are innumerable. For those inter-ested in the results and recommendations from the different studies, the study reports and the base paper summarising the results provide useful reading.

The intention here is not to analyse the results of the preparatory phase of the project in relation to the objectives of the same. Instead, I will try to present some lessons which relate to the biodiversity aspects of the project. My understanding of the process and of lessons learnt are based largely on my own involvement in the biodiversity study of the preparatory project phase (Singh et al., 1999). This means inter alia that as an actor in the first project phase I am biased, and I carry my biases with me into this case study report.

In order to attempt a meaningful analysis, the lessons presented in this case study may be separated into three levels:

- lessons related to development of the project concept;
- lessons learnt in preparation of the project, i. e. major results and recommendations from the studies in the project preparation phase; and
- lessons learnt regarding the project development process.

In the following, I will try to summarise some lessons learnt relating to biodiversity aspects on these three levels.

Lessons Related to Development of the Project Concept

Sida supported a Social Forestry Programme in Orissa between 1983 and 1996. This programme covered all of Orissa and focussed on plantation of village forests on land NOT classified as forests, i e land outside the normal domain of OFD. The major costs of the programme were staff (in a separate Social Forestry wing of OFD created for the programme) and labour for tree planting in village woodlots, weeding and guarding plantations etc. Through creating seasonal employment opportunities, the programme did contribute towards poverty alleviation. It was also relatively successful in establishing the plantations. However, the social objective of the programme, to create a sustainable income for the poorer sections of society through tree growing, was never realised to any larger extent (Ingevall, 1997; Holm & Nilsson, 1997). The Village Forest Committees which were created did not realise their actual ownership of the plantations, and distribution of benefits from these were not equitable.

The tree plantations in the Social Forestry Programme were almost exclusively monocultures of exotic tree species and hence with low natural biodiversity and production of few non-timber forest products beside fuel in the form of wood and dry leaves.

Based on the various difficulties and shortcomings of the Social Forestry Programme, Sida's support came to an end in 1996. Sida's negotiations with Orissa regarding a new forestry project has been a long process, and Sida has played an active role in supporting change within OFD. This has been the case also in the re-orientation from plantation forestry based on monocultures in the social forestry project to the focus on natural regeneration of forests in the PMDFO project. Conservation and sustainable use of biodiversity are integral to this focus, and so is a focus on equitable sharing of benefits from the use of biodiversity. All three major objectives of the Convention on Biologi-cal Diversity have thus been actively addressed by Sida in the dialogue with OFD (Ingevall, 1997).

Ultimately, Sida's support to the project is aimed at poverty allevia-tion, which is one of the main goals of Swedish development coopera-tion. In this project, the biodiversity objectives are intimately linked to the overall aim of poverty alleviation.

Sida's active role in supporting a focus within OFD on participatory management of naturally regenerating forests has been a key factor in the development of the project concept. Sida's active role was possible because of the professional skills of Sida staff in relation to sustainable and participatory natural resource management, a long history of cooperation with OFD and other actors in Orissa, presence in India, knowledge of national and state level policies and legislation of relevance in the project context, and persistence in the dialogue with OFD.

The key role of the donor in the dialogue with the cooperation partner is an important lesson in this case study.

Major Lessons from Studies in the Project Preparation Phase

The preparation of the PMDFO project was organised as a series of studies undertaken on issues of relevance for the project. A summary of lessons learnt from these studies, as presented in the project base paper. The studies did not cover issues related to reorganisation of OFD. The following is a brief summary of the major lessons which relate to the biodiversity aspects of the project.

Roles of Communities in Forest Management

In accordance with the national and state level forest policies, the broad objectives for management of CFM/JFM areas in Orissa should be: rehabilitation of degraded forests through natural regeneration, improving productivity particularly of non-timber forest products (NTFPs) through appropriate measures including biodiversity conservation, sustainable management and sustainable use to supply forest products to meet village communities' requirements. Within this broad framework, area-specific management objectives will be developed by each village level organisation (VLO).

The management objectives need to include an ecosystem perspec-tive, where issues related to e.g.fodder production and improvement of pastures both within and outside the protected forests may need to be addressed.

Participatory management systems need to be developed. Participa-tory forest management will include a baseline survey, a planning process using PRA methodology, and participatory monitoring and evaluation. Planning and monitoring will address managerial, finan-cial, slvicultural, social and biodiversity issues.

Policy and Legal Framework

The regulations and policies concerning NTFPs do not sufficiently recognise the rights of forest dwelling communities with respect to collection, processing and marketing.

Orissa government needs to carry out an internal review of policy and laws relating to NTFPs in order to provide clear definition of NTFPs, ensure transparent management operations in relation to collection, processing and marketing of NTFPs, sharing of revenue generated, institutional arrangements etc. The review also needs to ensure compliance with National Forest Policy and national conserva-tion guidelines.

The existing practice of monopoly leases in NTFP trade and process-ing should be discontinued. Leases could be given to a number of buyers including co-operatives, non-profit making societies, joint sector companies, VLOs and their federations. Formation of primary collectors' institutions could be helpful. The responsibility of primary collection, storage, and minor processing of NTFPs could be en-trusted to VLOs. Higher level processing of certain NTFPs could be allowed on experimental basis to capable VLOs.

The current ban on felling means locking the resource without its sustainable use. It is likely to negate the benefits of CFM/JFM in the state, and it should eventually be lifted. However, this would require the development of regulatory and monitoring mechanisms, to ensure (1) designationa and continued protection of ecologically important areas; (2) ecologically sustainable logging practices and (3) a system for fair sharing of benefits, where the private sector does not dominate the scene with minor benefits to the local communitites.

Needs for Training and Improved Knowledge

Increased production and collection of NTFPs will depend on partici-patory forest management systems where the forests are managed for production of both timber and NTFPs. Management systems that enable sustainable production and collection of NTFPs is generally expected to contribute to restoration of forests rich in biodiversity, which will be ecologically sustainable and hence contribute to sustain-able livelihoods of forest-dependent communities. However, produc-tion of some major NTFPs (e.g. kendu leaf, sal leaves and sal seed) may lead to simplified and eventually degraded ecosystems. A better understanding is needed of how best to optimise production of these NTFPs.

Appropriate training for both VLO members and OFD staff is important. Topics for training concerning silviculture and manage-ment of CFM/JFM areas may include i) importance of biodiversity in forest management, ii) natural regeneration survey and simple inventory of forest resource (number of trees and basal area per hectare), iii) sustainable silviculture and management practices for CFM/JFM areas, iv) management plan preparation, v) sustainable harvesting, use and marketing of products from CFM/JFM areas, and vi) community organisation and functioning of VLOs.-Preparation of a field manual on silviculture and management of CFM/JFM areas is essential for training as well as for extension support. Appropriate literature on silviculture and management of CFM/JFM areas needs to be published in Oriya for use by VLOs, OFD staff and NGOs.

The Biodiversity Study

The biodiversity study of the PMDFO project preparatory phase con-sulted members of 27 communities in 4 districts in Orissa. The bio-diversity aspects studied included:

- biodiversity and ecosystem functions in forests protected by local communities;
- collection and use of non-timber forest products (NTFPs) by different groups in the communities;
- study of social, cultural and spiritual values of biodiversity; and
- monitoring of biodiversity as an integral part of participatory forest management.

A summary of findings on the first three aspects has been presented earlier in this report. Below are some of the conclusions concerning biodiversity monitoring in participatory forest management. Biodiversity monitoring in participatory forest management Participatory forest management with integration of biodiversity aspects and for optimal yield of NTFPs would typically include management choices like selection of both timber and multi-purpose tree species in silvicultural management; avoidance of indiscriminate clearing of forest under-story, and keeping NTFP species in cases where under-story clearing is necessary (Scandiaconsult Natura AB/Asia Forest Network, 1998c).

Some NTFP collection, like sweeping of leaves from the forest floor (for fuelwood or to facilitate collection or e.g. sal seeds), harvesting of tubers and medicinal plants, and hunting, may well have negative effects on forest biodiversity and ecosystems if done indiscriminately. There is for instance evidence that sweeping of leaves has negative impact on development of forest soils in some protected forests.

In order to learn about the effects on biodiversity of management choices made, monitoring is needed. The best approach to biodiversity monitoring in participatory forest management, based on consultations with 27 communities, was found to be simple inventories done as transect walks through protected forest areas, documenting plant species and arthropod groups, complemented by an enumeration by the community of plant and animal species known to live in the forests, and trends (increasing or decreasing in numbers) for key species.

The biodiversity-NTFP study recommended that this simple biodiversity monitoring be integrated with the communities' participa-tory forest management planning as recommended by the study on participatory monitoring and evaluation of forest management (Balasubramanian, 1998). Biodiversity monitoring should thus be done by the community for its own needs. Detailed design of biodiversity monitoring can hence only be done by the communities themselves. This should form part of the implementation phase of PMDFO.

Lessons Learnt Related to the Project Development Process

Most of the studies carried out in the preparatory phase of PMDFO were done in consultation with local communities. The participatory approach was planned for at the inception of the preparatory phase. Consultations between local communities, NGOs and OFD in the form of seminars etc. were also planned in order to discuss the outcome of the different studies and constitute a basis for further project planning. Consultations were held between OFD and the major NGOs active in supporting CFM/JFM in Orissa in the form of a number of seminars during the preparatory phase of the project. This provided useful opportunities for exchange of views and sharing lessons regarding CFM/JFM between OFD and the NGOs.

Efforts were made to integrate the work of the different studies. For instance, the consultants carrying out the biodiversity-NTFP study worked actively to integrate this with the other studies. This was done by consultations with the consultants responsible for the other studies and by efforts to integrate preliminary results of other studies into the biodiversity-NTFP study and vice versa. Some of the field work under-taken in the biodiversity-NTFP study was done jointly with the study of silviculture and forest management. This was beneficial for mutual learning about relationships between biodiversity, silviculture and forest management.

However, consultations which include local communities and their federations regarding the outcome of the studies have not as yet been carried out. Without such consultations, the results and recommenda-tions from the preparatory phase can only be seen as preliminary. One of the factors restricting the level of consultation in the review stage of project preparation was shortage of time in the project planning process. This particularly restricted the possibilities to provide informa-tion to communities in the form of study outcomes and proposals as well as the opportunity of communities to comment on these. It was origi-nally planned to translate the study reports and proposals into Oriya. This, however, was not done due to shortage of time.

A major lesson learnt here is that a process for participatory review should be planned at the outset which guarantees that the different interest groups have opportunity and time to present their views on the outcome, and that their views are taken into account. To enable this, time must be set aside for participatory review at the end of the process in the original time plan (and any delays should not be allowed to en-croach on the time allowed for participatory review); a clear process for participatory review must be agreed on and adhered to; and the neces-sary documents need to be translated into the local language and distrib-uted in a timely fashion to the relevant interest groups. Since a participatory review was not held during the preparatory phase of the project, it could form part of the preparation for a possible continuation of the project.

Different views Revealed

In the studies, one interesting observation from the joint field work of the biodiversity-NTFP team and the silviculture/forest management team was that during the field visits, the dialogue of the two teams with community members yielded very different views on forest management. For example, when walking through a forest area protected by the community and discussing with community members, the silviculture/forest management team got the clear impression that ṣ 1 trees would be favoured in a thinning operation, while wild fruit tree species would be felled. Conversely, the biodiversity team were told that sal trees and wild fruit trees were equally important in forest management.

This points towards the need for great care in interpretations of results from brief visits to local communities, and towards the need for longer term interaction, where both parties get to know each other and find space for open dialogue. It also points towards the need for dialogue with members from different groups of the same community, since needs and interests necessarily differ between these. One reason for the different views disclosed in the two groups is probably the desire of the community to please the visiting team and fulfil the team members' expectations, in accordance with the focus and emphasis they provide in the brief interaction with the community.

In spite of these shortcomings, the NGO and community level consulta-tions in the PMDFO preparatory phase helped to facilitate the formation of the state-level forum of village level communities in Orissa, which was formed in March 1999. The separate support from Sida to NGOs supporting participatory forest management in Orissa also facilitated this process.

The main objective of the Odisha Jungle Mancha, or Orissa Forest Forum (OFF), is to act as a pressure group for formulation of a pro-people forest policy favouring the interest of forest protecting and man-aging village communities (Pal, 2000). OFF is now recognised by the Orissa Government as an advisor and acts as a bridge between the communities and the State. OFF's lobbying has enabled the formulation of Orissa's new NTFP Policy. Announced during 2000, the NTFP Policy gives rights over NTFPs to Panchayats, and what was previously a monopoly trade is now decentralised, with a minimum price back-up from the State. The communities can now freely trade and undertake value addition of NTFPs.

Although this development was only indirectly a result of the PMDFO preparatory phase, it is nevertheless a major achievement which the PMDFO process helped facilitate.

Environmental Impact Assessment of the Project Document

As agreed in the planning of the preparatory phase of the project, and in accordance with Sida's EIA Guidelines (Sida, 1998a), a brief EIA was done for OFD of the project document of May, 1999. The EIA was done by one of the consultants in the biodiversity study (Marie Byström) as a desk study of the project document. The EIA thus built on the knowl-edge gained in the biodiversity study and on previous consultations with local communities and consultants carrying out the other studies. In one sense, the environmental aspects were integrated in the planning process, and the EIA served the purpose of checking that the project document incorporated all relevant lessons learned in the planning process.

There was however no feedback process to ensure that OFD reviewed the EIA. In future planning processes, this should be ensured.

BIODIVERSITY ASPECTS OF THE PMDFO PROJECT IN RELATION TO NATIONAL AND INTERNATIONAL AGREEMENTS AND POLICY DEVELOPMENT

While the national and state-level forest policies of Orissa may be of more immediate importance for forest-dwelling people in Orissa, it may also be useful to see how the PMDFO project relates to international agreements related to biodiversity. India is a party to the Convention on Biological Diversity (CBD) and a member of the World Trade Organisation (WTO) and hence a party to the WTO agreement on trade-related intellectual property rights (TRIPs).

The Convention on Biological Diversity (CBD)

The three overall objectives of the Convention on Biological Diversity (CBD) are:

- the conservation of biological diversity
- the sustainable use of its components
- the fair and equitable sharing of the benefits arising out of the utilization of genetic resources.

The parties to the CBD are requested to act towards the fulfilment of these objectives at the national and local level and through international cooperation. Both the Indian national forestry legislation and the Orissa state level forestry legislation referred to above address all three major objectives of the CBD to some extent. The new NTFP Policy in Orissa (Gov of Orissa, 2000) will substantially increase the rights of local communities to process and market NTFPs, and local communities will hence hopefully share a larger proportion of the benefits from the use of NTFPs, which may at least partly be defined as genetic resources.

An Indian national biological diversity law, which more directly addresses the objectives of the CBD, is in the process of legislation. Among the features of the bill are the proposal of People's Biodiversity Registers for the documentation of local biodiversity knowledge, and of a National Biodiversity, State Biodiversity Boards and local Biodiversity Management Committees to regulate access to biodiversity and sharing of benefits. The bill has been developed in a participatory process. Improvements have been suggested by e.g. Madhav Gadgil et al. (Gadgil et al. 2000) to further focus the biodiversity law on conservation, sustain-able use and benefit sharing of biodiversity, and to strengthen the provi-sions for local management. The biodiversity bill, when passed, may give new opportunities to local communities in terms of sharing the benefits of commercial uses of biodiversity provided through their knowledge. In addition, the People's Biodiversity Registers may assist local communi-ties to strengthen their knowledge on local biodiversity and may also help communities develop management plans, where sustainable use of biodiversity is an integral part.

The PMDFO project can be seen as one of India's activities towards the fulfilment of the objectives of the CBD. The planned project specifi-cally addresses the objectives/obligations of e.g. CBD articles 5 (interna-tional cooperation), 6 (b) (integration of conservation and sustainble use of biodiversity into relevant sectoral or cross-sectoral plans, programmes and policies), 8 (f), (i), (j) (rehabilitation and restoration of degraded ecosystems, provision of conditions to enable conservation and sustain-able use of biodiversity, and maintenance of knowledge, innovations and practices of local communities and equitable sharing of benefits from these), and article 10 (b), (c), and (d) (adoption of measures relating to the use of biological resources to avoid or minimise adverse impacts on biological diversity, protection and encouragement of customary use of biological resources in accordance with traditional cultural practices that are compatible with conservation or sustainable use, and support to local populations to develop and implement remedial action in degraded areas where biological diversity has been reduced).

On Sida's side, cooperation in the development of PMDFO supports fulfilment of Sweden's obligations in e.g. CBD articles 5 (international cooperation), 18 (technical and scientific cooperation, in particular 18.2.: cooperation in development and implementation of national policies, human resources development and institution building, and 18.4. encour-agement and development of methods of cooperation for the use of technologies, including traditional technologies, in pursuance of the objectives of the CBD), and article 20.3. (provision of financial resources related to the implementation of the CBD through bilateral channels).

TRIPs

WTO's agreement on trade-related intellectual property rights (TRIPS) sets minimum requirements for the patentability of life forms and hence relates to access and benefit-sharing of biodiverstiy. Article 27.3 (b) of TRIPs reads: Members may also exclude from patentability; (b) plants and animals other than microorganisms, and essentially biological processes for the production of plants and animals other than non-biological and micro-biological processes. However, Members shall provide for the protection of plant varieties either by patents or by an effective sui generis system or by any combination thereof. The provisions of this sub-paragraph shall be reviewed four years after the entry into force of the WTO agreement.

Although it is very difficult to quantify the effects of this request in TRIPs for intellectual property rights on certain life forms, it is beyond doubt that the extension of intellectual property rights to life forms have direct consequences for the distribution of benefits from genetic resource use. In particular, they contribute to the present structural change toward privatisation and concentration of biological research. While this may not directly affect local communities in Orissa, the TRIPs provisions may have a long term impact on their shares of benefits from use of genetic resources.

The extension of IPRs to life in industrialised countries, which pre-ceded TRIPs, has largely taken place without a political level discussion of the proper balance between the need to stimulate technological development and other societal objectives. The balance created by present patent practice in developed countries, and codified in the TRIPs agreement, is now broadly questioned by governments and NGOs in developing countries, including India, from an ethical and socio-eco-nomic perspective. It would be advisable for developed countries to acknowledge this fact and return to a renewed consideration of their standpoints from this insufficiently explored angle. India is among the countries that have pointed to a potential conflict between their commitments under the CBD and under TRIPs, as the CBD stipulates that IPRs must be supportive of the Convention objec-tives, while TRIPs lacks reference to either conservation, sustainable use, or fair and equitable benefit sharing.

Among the concrete proposals given to WTO from developing coun-tries are the withdraw of the requirement of IPRs on life forms alto-gether, direct inclusion of some of the CBD objectives in the TRIPs text, and amending the text with explicit provisions for community rights protection. This would strengthen India's power to implement the benefit-sharing provisions in the proposed biodiversity law in an interna-tional perspective. Ultimately, this could benefit local communities in their forest management in Orissa and would hence support the objectives of the project PMDFO.

RELATIONS TO EC BDP'S GUIDING PRINCIPLES'

As part of the "Strategic Framework" developed for the EC/ DFID/ IUCN Biodiversity in Development Project, seven guiding principles have been developed (BDP, 2000).

BDP Guiding principles to ensure that all development cooperation actions are sustainable and effective, and give biodiversity proper consideration

1. Adopt an ecosystem perspective and multi-sectoral approach to development pro-grammes (taking account of impacts on adjacent and down-stream areas).
2. Ensure/encourage full stakeholder participation, including partnerships with civil society, government and private sector.
3. Ensure that development cooperation projects and programmes are consistent with the wider policy framework, and/or changes are made for supportive policies and laws.
4. Ensure that institutional arrangements are effective, transparent, accountable, inclusive and responsive.

5. Promote fair and equitable sharing of costs and benefits from biodiversity conservation and sustainable use, at local and national and international levels.
6. Provide and use accurate, multi-disciplinary information, which is both accessible to and understood by all stakeholders.
7. Development cooperation investments must be sensitive to, and complement local/ national structures, processes and capacities.

Below, brief comments are given on (A) findings in the case study as they relate to the Guiding Principles, and (B) the relevance of the BDP Guid-ing Principles to mainstreaming of biodiversity in development coopera-tion as judged from this case study. (The Guiding Principles were not developed at the time of preparation of the PMDFO project.)

A. Brief comments on findings in the case study as they relate to the Best Practice Principles:

1. Adopt an ecosystem perspective and multi-sectoral approach to development programmes (taking account of impacts on adjacent and down-stream areas).

   Comment: An ecosystem perspective was part of the project approach. The notion of positive impacts on adjacent and down-stream areas was an integral part of the motivation of many communities to protect forest areas, according to the biodiversity study.

2. Ensure/encourage full stakeholder participation, including partner-ships with civil society, government and private sector. Comment: Full stakeholder participation was aimed for in the project. The work in most of the different studies was done in consultation with local communities, and OFD had consultations with several NGOs during the preparatory phase. However, although working with participatory processes, the first phase had some shortcomings in relation to this principle. This was the case in particular in the review of results from the preparatory phase. To what extent has the private sector

been involved in the studies? This sector and its role in sustainable forestry management are not much mentioned in the report.

3. Ensure that development cooperation projects and programmes are consistent with the wider policy framework, and/or changes are made for supportive policies and laws.

   Comment: This was one of the guiding principles in the development of the project concept. The principle was also addressed by several of the studies, in the call for review of policies and laws related to NTFP production and marketing.

4. Ensure that institutional arrangements are effective, transparent, accountable, inclusive and responsive.

   Comment: The institutional arrangement of the project itself, with the project coordinator and Team Leader from the consultant, Ulf Öhman, stationed at OFD headquarters in Bhubaneswar and work ing very closely with a group of officers at OFD, met all these princi-ples to an unusually high degree. This is likely to in itself stimulate OFDs work at reorganisation of the Forest Department in line with the above principles.

5. Promote fair and equitable sharing of costs and benefits from biodiversity conservation and sustainable use, at local and national and international levels.

   Comment: The project "Capacity Building for Participatory Manage-ment of Degraded Forests in Orissa, India" aims at promoting sus-tainable and community based management and use of forests in Orissa. This overall aim includes the promotion of fair and equitable sharing of benefits from biodiversity conservation and sustainable use at state level in Orissa. This principle was also clearly addressed by nearly all of the studies in the preparatory phase of the project.

6. Provide and use accurate, multi-disciplinary information, which is both accessible to and understood by all stakeholders.

Comment: Although working with participatory processes, the first phase had some shortcomings in relation to this principle, as discussed in section 6.3. This was the case in particular in the review of results from the preparatory phase. This study recommends for future work in the project (1) translation of relevant documents into the local language, and (2) timely distribution of information to allow review by all relevant stakeholders.

7. Development cooperation investments must be sensitive to, and complement local/national structures, processes and capacities.

   Comment: This was a very central guiding principle in the development of the project concept. In contrast, the earlier Sida-supported Social Forestry Project in Orissa did not have sufficient sensitivity to local and national structures, processes and capacities. This was one reason for Sida to end the support to this project.

B. Comment on the relevance of the BDP Best Practice Principles to mainstreaming of biodiversity in development cooperation:

The best practice principles are excellent guiding principles for develop-ment cooperation. The fifth principle, "Promote fair and equitable sharing of costs and benefits from biodiversity conservation and sustain-able use, at local and national and international levels" is directly related to mainstreaming of biodiversity in development projects. All the other principles should be guiding principles for development cooperation in general, and the fifth principle should be promoted not only in "biodiversity projects" but also in work towards the mainstreaming of biodiversity aspects more widely in development cooperation in the natural resources management sector.

CONCLUSIONS AND RECOMMENDATIONS FOR BIODIVERSITY MAINSTREAMING AT SIDA

There are no shortcuts to integration of biodiversity aspects into natural resource management projects/programmes.

Integration, or mainstreaming, will depend on the understanding by the actors involved of the relevance of biodiversity issues in the project/pro-gramme context.

The dialogue between Sida and the cooperation partner is a key factor. A genuine dialogue requires time and concerted efforts on both sides.

Among the most important factors for integration of biodiversity issues in this project was the professional competence of Sida staff, as well as intimate knowledge of the situation in the country/area of cooperation, including knowledge of the policy/ legal context in relation to biodiversity issues and related socio-economic issues.

Hence, it seems important that Sida staff has sufficient time to (1) follow the international discourse and policy development in biodiversity-related areas; (2) acquire a high level of knowledge of the country-specific situation, and (3) undertake a genuine dialogue with the cooperation partners, with sufficient time for the actors concerned to reflect on and come back to crucial issues.

In project planning/implementation, clear processes for participatory planning, monitoring and review should be agreed upon early in the planning stage and adhered to. Of particular importance is the need to ensure that all interest groups are able to participate to sufficient degrees.

Monitoring of biodiversity as part of natural resource management should be:

- as simple as possible in order to be cost/effective, manageable and replicable;
- based on local knowledge;
- an integral part of local management systems.

Sida's strategic priorities in relation to the CBD (Sida, 1998b) were found to be relevant in this case study. Sida's guidelines for support to biodiversity from 1994 were also relevant (Sida, 1994). The focus of the project studies is directly related to two of Sida's three strategic priorities for biodiversity, namely support to:

- work to respect, maintain and develop knowledge on the conservation and sustainable use of biological diversity in local communities and indigenous populations including support for strengthened local control over the use of biological resources.
- the protection and sustainable use of biological diversity in areas which are utilised by human beings including agriculture, forestry and fisheries. The focus should lie on mechanisms which make it possible to continue to maintain the sustainable use of biological diversity at higher levels of production.

No need to revise Sida's strategic priorities was identified in this case study.

In Sida's Guidelines for EIA in development cooperation (Sida, 1998a), the following questions could be added in the biodiversity section in the relevant checklists:

"Will the project facilitiate or work against the development of fair and equitable sharing of benefits from the use of biological diver-sity?"

"Does the project involve activities which require the prior in-formed consent of local communities or indigenous peoples? If so, have these groups been informed? Have they given their consent to the planned activities? Are the plans on mutually agreed terms?"

Development of processes for the integration of EIA in project planning may be needed. Integration of EIA in the planning process depends on the degree of ownership of the EIA felt by the actors involved. Development of a sense of ownership of the EIA requires understanding of the relevance of EIA as a useful tool in the planning process. Sharing of lessons learnt between projects will be important in this context.

The "Guiding Principles" of the EC/IUCN/DFID Biodiversity in Development Project (BDP, 2000) were found to be relevant to the project studied, and they should be used as guiding principles not only for mainstreaming of biodiversity but for development cooperation in general.

In "Sida and the Convention on Biodiversity" Sida assumes the responsi-bility for mainstreaming aspects of biodiversity into all programmes, starting-with the Department for Natural Resources and the Environ-ment (NATUR).

The mainstreaming will focus on analysing the conse-quences for biodiversity of the programme/project and on making stakeholders, women as well as men of different ages, aware of the importance of biodiversity on all levels from gene, population, species, functions and ecosystems, both wild and cultivated biodiversity.

Mainstreaming of biodiversity will have more relevance in some pro-grammes/projects of development cooperation than others. The reason for the selection of NATUR as the first department at Sida in the mainstreaming process is that NATUR is responsible for contributions to agriculture, forestry, fishing, etc which depend on biodiversity and have a direct impact on biological diversity.

The objective (according to the Sida-memo on "Mainstreaming of Biodiversity at Sida-phase 1") of this work is:

> *"that consequences for biodiversity are analysed in the project identification, planning process and follow-up of all programmes and projects supported by Sida-DNRE, as part of the EIA, to minimise negative effects and also point out positive impacts for biodiversity"*

A simultaneous processes take place with the same purpose of main-streaming biodiversity into development cooperation programmes, e.g. the Biodiversity in Development Project (BDP) undertaken by EC in collaboration with DFID and IUCN. A coordination with the process in the BDP will take place.

Purpose and Scope

As part of the process of mainstreaming biodiversity aspects into prepa-ration, implementation and monitoring of Sida-DNREs natural re-sources management and rural development programmes, a number of case studies will be undertaken. These will form the base for developing hands-on methods and guidelines on biodiversity mainstreaming for primarily Programme Officers at Sida-DNRE. Sida will obtain inputs on:

- methods for how aspects -of biodiversity can be mainstreamed into both the assessment phase of a project/ programme and on-going projects/programmes
- how to monitor aspects of biodiversity in the programmes/ projects (through environmental indicators etc).

Tasks

The task include:

Undertaking Case Studies in Connection to three to four Sida-DNRE Bilateral Programmes

The case studies shall describe and analyse to what extent biodiversity aspects have been considered within the programme context. Two Sida-programmes have so far been identified as suitable for case studies: The Joint Forest Management Programme in Orissa, India, presently under preparation, and the Mountain Rural Development Programme (MRDP) in Vietnam. One to two more case studies remain to be identi-fied from Africa, and possible suggestion include the Region 3-pro-gramme in Ethiopia, LAMP in Tanzania, the agricultural sector pro-gramme in Zambia, the planned support to the agriculture sector in Mozambique. Other options are the Lake Victoria-initiative, or the water-related activities in southern Africa.

The case studies shall cover the following questions:

Background

Background and history of the project/programme, including descrip-tion of the programme's various phases and the national development context to be able to understand in which context biodiversity is to be mainstreamed. This includes a description of policy and legal framework for the country. Which treaties, binding and non-binding, have been adopted such as:

- World Intellectual Property Organisation's (WIPO's) Union for the Protection of New Varieties of Plants (UPOV), binding

- World Trade Organisation's (WTO's) Trade-Related Aspects of Intellectual Property Rights (TRIPs), binding
- Convention on Biological Diversity (CBD), binding
- Food and Agriculture Organisation's (FAO's) International Undertak-ing on Plant Genetic Resources and the Global Plan of Action for the Conservation and Sustainable Utilisation of Plant Genetic Resources for Food and Agriculture
- Rio-declaration including the forest principles, non-binding recom-mendations
- Agenda 21, non-binding Action Plan
- Convention on International Trade in Endangered Species of Wild Fauna and Flora (CITES), binding
- RAMSAR-Convention on Wetlands, binding
- and other treaties in relevance for the programme.

It also includes a description of which donors that exist in the country.

Project/Programme Description

Brief description of the present set-up of the programme:

(a) description of the aims and activities of the project

(b) geographical area

(c) socio-economic and cultural context

(d) stakeholder analysis including ethnic and gender relations, and also an assessment of stakeholders that were "excluded" from the project activities

(e) institutional set-up including policy framework on biodiversity

(f) programme activities.

Problem Analysis

Analysis of biodiversity aspects within the programme, both historically (since inception of programme) and presently. The analysis should:

(a) be made using the following documents as analytical tools:

- "Biological Diversity-guidelines for Sida support for the sustainable use and conservation of biodiversity", Sida, 1994
- "Sida and the Convention on Biological Diversity", Sida, 1998
- "Guidelines for Environmental Impact Analyses in development cooperation", Sida, 1998
- Sida's four action programmes
- "Strategic Framework" developed for the Biodiversity in Development Project (to be able to influence the BDP-process).

(b) explicitly consider and comment on the full range of biodiversity issues in the project area, (e.g. evidence of irreversible losses, opportu-nities for improved use of biodiversity for human development, etc), to what extent biodiversity has been "mainstreamed", which biodiversity aspects that have been considered, and the relevance of these to the different stakeholders, how the aspects of biodiversity has been handled and from which stakeholders,

(c) identify key constraints to improved biodiversity management and improved human development, and assumptions made about the links between the two, within the context of the project aims and activities

(d) include assessment of whether project/programme successes have been, or are, sustainable (considering mainly aspects in connection to biodiversity use) in the long term (i.e. without further project input),

(e) explicitly consider and comment on methods used for mainstreaming aspects of biodiversity into the project/programme, past and present monitoring of biodiversity aspects and tools to follow-up these aspects (environmental/biodiversity indicators).

Particular attention should be paid to the analysis of the projects impact on ecosystem services related to biodiversity, as well as the linkages between biodiversity and sustainable livelihood, culture and social relations (including ethnic, gender and age relations) and political and socio-economic structures. Also gene policy questions (bioprospecting and eroison of domestic species, plants and animals, genetic diversity) should be analysed.

Institutional Resources

Identification of and consultation with possible other institutions of importance, shall be done, that could be a resource for the programme (regarding biodiversity).

Lessons Learned for the Programme

The lessons learned from the mainstreaming of biodiversity for the project/programme planning and implementation, including field methods, shall be summarised and discussed.

Lessons Learned for the Sida Mainstreaming Process

The lessons learned from the mainstreaming of biodiversity for project/ programme planning and implementation, including field methods, shall be summarised and discussed.

Conclusions and Recommendations for the Programme

Conclusions and recommendations for the programme:

(a) on methods for biodiversity mainstreaming within the programme

(b) on follow-up mechanisms (monitoring including possible indicators).

Conclusions and Recommendations to SIDA

Conclusions and recommendations to Sida regarding:

(a) improvements that could be made in the project/ programme plan-ning process and implementation of projects/programmes

(b) improvement of the analytical tools and guidelines, see c) above, for mainstreaming biodiversity aspects within Sida-DNRE's programmes.

(c) methods for how aspects of biodiversity can be mainstreamed into projects/programmes.

(d) how to monitor aspects of biodiversity in the programmes/projects (through environmental indicators etc).

Presentation of Case Studies in Seminar/workshop

The case studies will be presented in a seminar/workshop at the Sida office in Stockholm in January 2000, and when possible in the country where the case study has been taken place. The case studies will also, if possible, be presented at the BDP regional workshops on case studies: Cameroon (28/6-2/7/99); Sri Lanka (24/7-30/7/99); Botswana (6-11/9/99); Peru (27/9-1/10/99).

METHODS

The mainstreaming of aspects of biological diversity is of course de-pendent on the interest of the "owner" of the project/programme in extending the focus of the project/programme. The mainstreaming should be initiated in close connection with regular reviews of Sida's support.

The consultants performing the case studies will be responsible for developing specific methods for each case study, as a contribution to lessons learned, in cooperation with the recipient partner. The consult-ants shall try to coordinate the activities in the case studies with the BDP, see also Terms of Reference for BDP case studies, appendix 2. The work on the case studies can be done in the following way:

(1) Discussion and planning in consultation with programme officers concerned.

(2) Planning of each study in consultation with the partner in coopera-tion, institutions of importance for the issue in the recipient country and consultants responsible for the implementation of each pro-gramme.

(3) Implementation in consultation with, and preferably together with, the partner in cooperation and institutions of importance for the issue in the recipient country, if possible in the field.

(4) Review of results in seminar form (and in a written report) in connec-tion with annual reviews or suchlike, and also with the staff at Sida-DNRE. Discussion of results and recommendations for each pro-gramme.

(5) The follow-up of the first mainstreaming studies should take place after approximately one year, in connection with the normal follow-up of the programme. At this point in time the methods can be evaluated and further inputs to the analytical tools and guidelines can be made. From our own results and from the BDP results, further recommendations can be made on how the work of mainstreaming aspects of biodiversity can be continued in other projects/pro-grammes in the natural resources sector.

TIME FRAME

The case studies, for mainstreaming of Biodiversity at Sida phase 1, DNRE, shall all be completed within the year 1999, although follow-up of the case studies and further lessons learned from them will take place during year 2000 and beyond. (6. Specifics regarding the Vietnamese case study, omitted here)

SPECIFICS REGARDING THE ORISSA CASE STUDY

The project "Capacity Building for Participatory Management of Degraded Forests in Orissa, India" aims at developing the concept and understanding of Participatory Forest Management (PFM) that includes both joint and community forest management as a potential for sustain-able use of forest resources in the state of Orissa.

The understanding of PFM should be based on experience from Orissa Forest Department (OFD) and from the experiences of local initiatives to protect forest areas for sustainable production of timber and non-timber forest products (NTFPs).

The first phase of the project (Dec 1997-May 1999) aims at devel-oping a background and capacity for a longer support in the second phase. The outcome of the first phase will be a proposal for continuation of the project, written by OFD but developed out of an understanding of PFM for sustainable utilisation of forest resources shared between OFD and NGOs and village level organisations (VLOs) active in the area.

The multiple goals of the first phase are:

1. Further elaboration of JFM and CFM concepts by exploring relation-ships between joint and community based forest management prac-tices;
2. Strengthening the capacity, including restructuring, of the Orissa Forest Department;
3. Knowledge about low cost methods for reforestation through natural regeneration;
4. Use of the information and learning experiences generated during this period for the preparation of the project document for phase II.

Purpose and Scope

The purpose of the biodiversity case study is to draw and present lessons learnt from the first phase of the project "Capacity Building for Partici-patory Management of Degraded Forests in Orissa, India", in particular regarding mechanisms for mainstreaming biodiversity aspects, which biodiversity aspects have been covered in the project and the relevance of these to the different stakeholders.

The study will present the format used in the project and proposed by the project for PRA-based village studies and monitoring on environment and biodiversity, including environmental and biodiversity indicators and with special emphasis on conclusions of relevance for other Sida-sup-ported programmes.

The study will also make conclusions and recommendations, based on the experience from the project "Capacity Building for Participatory Management of Degraded Forests in Orissa, India", regarding improve-ment of the analytical tools and guidelines for mainstreaming biodiversity aspects within Sida-DNREs programmes.

Methods

The study shall be undertaken through:

- studying the different reports from phase one of the project to draw lessons regarding to what extent biodiversity has been "mainstreamed", which biodiversity aspects that have been consid-ered, the mechanisms for mainstreaming biodiversity aspects, and the relevance of these to the different stakeholders;
- consultations with programme officers concerned at Sida-DNRE;
- consultation with relevant project staff;
- consultation with other institutions of importance in India;
- studying other background material of relevance for the case study;
- presenting lessons learnt from the project, in particular regarding mechanisms for mainstreaming biodiversity aspects and the relevance of these to the different stakeholders;
- presenting the format used in the project and proposed by the project for PRA-based village studies and monitoring on environment with specific emphasis on biodiversity and biodiversity indicators.

Time Frame and Reporting

A total of two weeks are required for the case study. Since field work for mainstreaming of biodiversity aspects in the project has already been done by the consultant, including consultations with stakeholders (repre-senting women and men from tribal groups, scheduled casts, open casts etc.) in twelve villages in Orissa, consultations with local and state level NGOs and with local and central OFD staff, no field work for this case study should be necessary. Necessary contacts in India should be possible to make through different forms of correspondence. A draft report will be presented by June 30, 1999. The consultant shall present the results in a seminar/workshop at the Sida office in Stockholm and also if possible in India.

REGARDING THE OTHER ONE TO TWO CASE STUDIES

The other one to two case studies are yet not elected but will be elected latest February 1999. They shall be finalised before December 1999.

Summary of project-specific lessons learnt from the studies in the pre-paratory phase of the project "Capacity Building for Participatory Management of Degraded Forests in Orissa, India"

In 1997, Orissa Forest Department embarked on preparation of the project titled "Capacity Building for Participatory Management of Degraded Forests in Orissa" with funding from Sida. Scandiaconsult Natura AB and Asian Forest Network assisted Orissa Forest Department in the project preparation phase, which took place during the period January 1998-May 1999.

The first phase of the project preparation was organised as a series of studies undertaken on issues of relevance for the project. A brief sum-mary of lessons learnt in the different studies is presented below. The lessons learnt have been synthesised in one summary, and they are not presented study-wise.

Forest Management

The management of CFM/JFM areas needs to be site specific.

No uniform set of silvicultural and management practices should be followed for all areas. Silviculture and management practices should suit the silvicultural requirements of the crop and should be adjusted to meet the objectives of management. The broad objectives for management of CFM/JFM areas should be: rehabilitation of degraded forests through natural regeneration, improving productivity particularly of NTFPs through appropriate measures including biodiversity conservation, sustainable management and sustainable use to supplement supplies of forest products to meet village communities' requirements. Within this broad framework, area-specific management objectives will be developed by each VLO.

Natural regeneration will be relied on to regenerate forests. No gap planting is recommended. The gaps and blank areas in CFM/JFM areas should be developed for grass production by cutting tall weeds and bushes not yielding any fodder. Fodder is an important requirement of village communities and development of such areas for grass production will be better than planting of trees. An efficient participatory learning and monitoring approach needs to be adopted not only to develop appropriate management practices, but also to ensure their proper implementation.

A baseline survey through which the status of the forest and socio-economic conditions of the village will be analysed through PRA before the preparation of the management plan.

Participatory monitoring and evaluation (PAME) should be done to review the progress of the forest management. It may be undertaken once in a year with participation of both VLO and OFD field staff. PAME would address managerial, financial, slvicultural, social and biodiversity issues.

Strong extension support to be provided by OFD ad NGOs is needed for proper management of CFM/JFM areas. Special training will be necessary for OFD field staff (see further below).

The current ban on felling (in other words locking the resource without its sustainable use) is likely to negate the benefits of CFM/JFM in the state. Its adverse effects include:

- lower productivity of the forest resource;
- increased incidence of illicit felling to meet villagers' requirements;
- reduced quality of growing stock as a result of illicit cutting of better stems and leaving the inferior ones as was observed during field visits, and
- mistrust between villagers and OFD because of the feeling gaining ground among the villagers that FD is interested in getting the JFM forests protected and rehabilitated and not in giving them the prom-ised forest products; such mistrust can be suicidal for JFM pro-gramme.

Sustainable use of NTFPs

The collection and use of NTPF provide an important part of subsist-ence and income, particularly for the tribal groups. NTFPs regularly collected for domestic consumption include fuelwood, mushrooms, medicinal plants, green leaf vegetables, fruits, tubers, resin, fibres, thatch-ing grass, and dyes. Important NTFPs collected and sold commercially include kendu leaves, siali leaves, sal leaves, sal seed, and mahua flowers. Present levels and methods of collection of NTFPs seem to be overall sustainable, with considerable potential for increased production and collection.

Increased production and collection of NTFPs will depend on partici-patory forest management systems where the forests are managed for production of both timber and NTFPs. Management systems that enable sustainable production and collection of NTFPs generally is expected to contribute to restoration of forests rich in biodiversity. However, produc-tion of some major NTFPs (e.g. kendu leaf, sal leaves and sal seed) may lead to simplified and eventually degraded ecosystems, if production is not carefully managed on a sustainable basis. A better understanding of how best to optimise production of these NTFPs is, therefore, important. The regulations and policies concerning NTFPs do not sufficiently recognise the rights of the forest dwelling communities with respect to collection, marketing and processing. While there is scope for a manifold increase of the production and value addition of NTFPs, it will depend on policy and legal changes.

The issues in NTFP trade relate to ownership rights over forest land and produce, accessibility, forms of disposal, marketing arrangements as well as procurement of raw material and price fixation. The existing practice of monopoly leases in NTFP trade and process-ing should be discontinued. Once the royalty and minimum support price etc. are decided and regulated, lease could be given to a number of buyers including co-operatives, non-profit making societies, joint sector companies, VFCs, VSSs and their federations. Formation of primary collectors' institutions could be helpful. The responsibility of primary collection, storage, minor processing etc. of NTFPs could be entrusted to VLOs. Higher level processing of certain NTFPs could be allowed on experimental basis to capable VLOs. Primary collectors/VLOs may be permitted to sell a major part of their finished forest products to ORMAS.

Legal/policy Issues

Orissa government needs to carry out an internal review of policy and laws relating to NTFPs in order to provide clear definition of NTFPs, ensure transparent management operations in relation to collection, processing and marketing of NTFPs, sharing of revenue generated, institutional arrangements etc. The review also needs to ensure compli-ance with National Forest Policy and national conservation guidelines. Policy constraints include:

- the provisions in JFM resolutions (1993 & 1996) regarding the roles of VLOs and FD need clarity to empower the VLOs to implement the programme and the FD to assist and facilitate implementation;
- usufruct sharing mechanism requires clarity, and major harvest needs a clear definition;
- a provision appears necessary to enable sale of products in excess of village subsistence needs, and appropriate marketing system needs to be developed to improve the returns to VLO members, particularly the poor, from CFM/JFM;
- harvesting trees should be the responsibility of the VLOs; the govern-ment resolution on JFM is not clear as to whether the trees to be removed in thinning are to be harvested by VLO members or by FD;
- ban on fellings in CFM/JFM areas needs to be removed to meet the requirements of management plans;
- CFM/JFM areas need to be taken out of the purview of working plans.

Training

Appropriate training for VLO members and OFD staff is important. Topics for training concerning silviculture and management of CFM/ JFM areas may include i) importance of biodiversity in forest manage-ment, ii) natural regeneration survey and inventory of forest resource (number of trees and basal area per hectare), iii) sustainable silviculture and management practices for CFM/JFM areas, iv) management plan preparation, v) sustainable harvesting, use and marketing of products from CFM/JFM areas, and vi) community organisation and functioning of VLOs.

Preparation of a field manual on silviculture and management of CFM/JFM areas is essential for training as well as for extension support. Appropriate literature on silviculture and management of CFM/JFM areas needs to be published in Oriya for use by VLOs, OFD staff and NGOs.

The training programme to be designed for phase II needs to inte-grate biodiversity aspects in all its parts. Training needs related to biodiversity include i) training of VSS and VLOs in participatory man-agement, monitoring and evaluation of biodiversity and NTFPs, ii) training of FD at central level, divisional level and local level in partici-patory management, monitoring and evaluation of biodiversity and NTFPs, and iii) training and skills development at village level for value addition of NTFPs, including packaging, storing, marketing, accounting and other management skills. Appropriate training for NGOs also should be arranged.

Training should be conducted as much as possible in the field. Training in skills development at village level for value addition and marketing of NTFPs needs to be developed together with local villagers and NTFP producers. Successful cases of local value-adding and market-ing of NTFPs may be used as examples in this training.

5

Valuation of the Mangrove Ecosystem

Due to shortage of time and fund, the project focuses on partial analysis, including cost-benefit analysis of planting, protection, thinning and extraction options, fishing and part of benefits rising from eco-tourism, while indirect-values or function and existence values will be initiate for further studies.

All marketable data available has been collected and collated, and the solicited by the intensive surveys in December 1999 and January 2000. A set of secondary data available in Management Boards of Can Gio Forestry Park, Statistic Office of District Can Gio and Management Boards of Protected Forests is being accessed, analyzed and aggregated by computerization. A new approach on economic valuation based on Total Economic Value (TEV) is introduced in the case study. The analysis framework of direct use values and non-use values of mangroves which is developed by using Stella 2 dealing with ecosystem services of mangroves in can Gio is tested. A detailed description of methods estimating the direct value of mangroves is described, including their limitation.

Based on Cost-Benefit Analysis, it is presented for each direct benefits/costs: planting, thinning, branching, nipa thatching, propagule supply, in-site and off-site fishing catch, aquaculture, fishing trades and travel cost analysis of eco-tourism. Some avoided costs / benefits is not estimated for shoreline protection, flood and storm mitigation and stock of water. However, the value of biodiversity conservation is evaluated by using the cost or annual investment from HCMC, as considered the existence value. It needs to be added and improved.

The results of cost benefit analysis (CBA) shows that it is of a partial nature, comparing planting, protection and thinning costs, with the direct benefits from extracted marketable products, and with the indirect benefits of the City' investment to restoration of the biosphere reserve. A benefit to cost (B/C) ratio in range of discount rates from 1 to 10 % is analyzed. For the B/C Ratio of direct benefits, the figures illustrates that most of the costs for planting from initial stage and protection and management system is less than the direct benefits. It can be explained that the limitation of thinning areas, over exploitation of fishing catch, water pollution, oil spill and others related is contributing the low direct benefits, but still higher than cost. However, this is also not meaning that local livelihood of people can be able to sustain in terms of basing the direct benefits.

Otherwise, a new regulation of banning any thinning activities from the City Administration has been launched in September last year and enforcement from early this year. According the regulation, a triple incentive will be given to households allocating forestry lands and protecting forests. However, a detailed response from communities has not been recorded. For the B/C Ratio of total benefits, the big different of benefits to costs is showing the vital role of investments from the National Government and City Administration to restore the reserve.

Based on the estimation of total economic value of mangrove restoration and rehabilitation, management options in dicision-making processes can be laid out from current to hypothetical scenarios in order to create a scene from unsustainable development up to sustainable one for resource use, like mangroves. Humain activities as driving forces effect to both socio-economic and natural processes.

The total economic value can be used as a tool in estimating the activities, especially the mass conversion of mangroves to other purposes. It was really an exchange of three kind of benefits (local, regional and global) in sustainable to one private benefit in short time. As an economic indicator, the total economic value may reflect the effects of policies in management and development in feedback estimation.

This is initial work to contribute the analysis framework needing an investment of time and funding, especially indirect use value, option and quasi-option values and existence value or non-use value as described by Barbier (1989, 1993, 1994) and Scodari (1990). The estimation of non-market values is a big challenge of environmental economists. However, valuation methods proposed by R. Costanza et al. 1997 on ecosystem services, natural capital and GNP are being applicable for this issue. It would be so kind to ask UNESCO/MAB to invest more time and funding for the work.

INTRODUCTION

The Can Gio Mangrove Biosphere Reserve has been designated for inclusion in the World Network of biosphere Reserve, by decision of the International Coordinating Council of the Program on Man and the Biosphere, in January 21, 2000.

Consisting of 77 mangrove species (35 true mangroves and 42 associates), the mangrove has contributed an economic and environmental importance in the coastal zone, as long as very vulnerable to impacts from natural and human driving forces. For almost population living in these areas, the mangrove provides not only valuable products from forests including timber, firewood, charcoal, tannin, food, medicinal but also the breeding ground for many species of marine organisms like shrimp, crab, fish, water and migratory birds as well as other economically terrestrial species of monkey, wild boars, boas etc.

It has been recognised by economists that the functions and services provided by mangroves, and wetlands in general, have positive economic value and that these are often ignored in the ongoing process of mangrove conversion (Barbier, 1993; Ruitenbeek, 1994; Turner, 1991; Swallow, 1994; Farber and Costanza, 1989).

The various functions and services provided by mangrove areas have been documented and appraised (e.g. Mitsch and Gosselink, 1994; Reimold, 1994; Lugo and Snedaker, 1974). Mangrove wetlands display the features of public goods in that their use is non-exclusive, and they are converted to other uses with private economic returns because the functions and services associated with them are undervalued. Identification of the functions and services and the incorporation of these into policy and the designation of property rights are therefore necessary first steps in promoting sustainable utilisation of such resources. The valuation of functions and services has often been expressed in a framework of the total economic value.

Critical issues in promoting the adoption of such schemes, and hence their ultimate sustainability, include the timing of the costs and benefits and the sensitivity of the economic appraisal of such schemes to discount rates and the limit of the benefits of rehabilitation compared to conservation of existing wetlands (reflected in the absence of option and existence values for such habitats). In the instance under analysis here, the benefits of direct use value of timber, non-timber products including fish, shrimp, crab and thatching, and other benefit for recreation, biosphere reserve as well is an important issue

THE STUDY SITE

The Biosphere Reserve is located on Can Gio District, about 65 km south of Ho Chi Minh

City with latitude: 10° 22'14"-10°

40'09" and longitude: 106°

46'12"-107°

00'59". It is 35 km

long from North to South and 30 km from East to West. During the two Indochina wars, almost all the mangroves in Can Gio were destroyed. After many years of herbicide spraying, the degraded land here still remains shattered bushy or bare land. A great effort of local people after wars is the rehabilitation of 21,000 ha of mangroves.

To date, Can Gio has become one of the most beautiful and extensive sites of rehabilitated mangroves in the world with high values of biodiversity, as seen in table 1.

Table : Mangrove areas and its management system up-dated in 1999

	Forestry Compartments Planted forests (ha)	Natural forests (ha)	Non-mangrove vegetation (ha)	Total (ha)
1.	644	294	292	1,229
2.	1,613	317	451	2,381
3.	550	463	154	1,166
4.	1,124	334	224	1,682
5.	719	187	228	1,134
6.	940	350	366	1,657
7.	496	231	444	1,172
8.	934	180	340	1,454
9.	859	407	180	1,446
10.	1,126	330	141	1,597
11.	657	289	422	1,367
12.	742	136	458	1,336
13.	880	201	802	1,882
14.	659	281	539	1,478
15.	866	401	674	1,941
16.	642	234	722	1,598
17.	1,164	378	691	2,233
18.	652	284	744	1,680
19.	480	295	472	1,247
20.	540	478	892	1,910
21.	641	169	794	1,604
22.	265	236	231	732
23.	2,474	74	201	2,748
24.	1,438	476	627	2,541
Total	21,104	7,026	11,087	39,217

Can Gio's topography is uneven. Its borders rise gradually to form the central hollow bottom, comprising 6-types of land topography (Un-flooded, Multi-year cycle flooded, Annual cycle flooded, Monthly cycle flooded, Daily cycle flooded and Accretion type).

Can Gio is situated in the typical tropical monsoonal zone with 2 distinctive seasons. The dry season lasts from November to the next May, the rainy season from the end of May to the end of October. The temperature is high, stable, the monthly mean temperature is 25.5-29.0° C. The daily average temperature amplitude is 5-70C. The insolation is 5-9 sunshine hours per day. The solar radiation is 10-14 kcal/cm2/month on an average. The radiation intensity does not vary much with different seasons.

The monsoons form different wave directions and coastal currents, directly influencing Can Gio's coast, leading to erosion and sea level rise. Compared to other areas in Ho Chi Minh City, the rainfall in Can Gio is lower and decreases gradually southwards. The rainy season begins on 20-25 May and ends around 25-31 October. The average annual rainfall is about 1,336mm. September sees the highest rainfall (300-400mm). The mean annual relative humidity is 80%. In the dry season, the air humidity during daytime is usually under 60%. The evaporation is 3.5-6mm/day and in March and April 7-8mm/day at the highest. It is 2.5-5.0 mm/day in the rainy months

Salinity: The survey data after 11 years (1977-1988) (Thanh, 1988) showed that in the dry season, highest salinity in March, April being 1.9-2.0% at Nha Be area, 2.6-3.0% near the sea. In the rainy season, the salinity is low; from August to October, the salinity in mangrove areas is only 0.4-0.8%. The average monthly salinity is 1.8%. Since 1988, the drainage of water from the reservoir of Tri An Hydroelectric Plant in the dry season has decreased the salinity in canals and rivulets of this area.

Given the prevailing circumstances in the coastal districts of Can Gio, and similar regions elsewhere, it is clear that mangrove rehabilitation can have a variety of benefits. In such situations, mangrove rehabilitation can provide income where households are often severely constrained in cash income sources, as well as bringing about environmental benefits in terms of productive assets and biodiversity resources. The following sections quantify an economic model of such an approach.

METHODOLOGY/APPROACH

The approach of total economic valuation is used in this work. Total Economic Value (TEV) is given by the sum of a number of components:

TEV = Direct-use value + Indirect-use value + Option value + Existence value

Direct use values include revenues from timber and values of other products such as thinning from Nipa palms. Indirect-use values or 'functional' values relate to the ecological functions performed by mangroves, such as global geochemical cycling, the protection of agricultural areas, and the provision of spawning grounds for fisheries. Option value, or more precisely, quasi-option value is the expected value of the information on the benefits of an asset, conditional on its preservation enabling an increase in the stock of knowledge relevant to the utilisation of the asset. A frequently evoked example of quasi-option value is associated with genetic resources: for example, the future value of pharmaceuticals developed from plant materials.

Existence value relates to the value of environmental assets irrespective of current or optional uses. Empirical measures of existence values based on donations to conservation organisations, or on the contingent valuation method, suggest that these can be a significant element in total economic value, especially in contexts where the asset has unique characteristics or cultural significance.

Although intuitively appealing, applications of Total Economic Value to particular environmental resources are few because of the problems of double counting if the elements are simply added and of quantification of the components. For example, although direct and indirect benefits of mangrove conservation can generally be assessed, the option and existence value of mangroves are difficult to determine within an economic analysis. Much research has been undertaken in this area, and some studies suggest that non-use values (option and existence) can be at least as large in magnitude as use values (see Walsh et al., 1984; Cummings and Harrison, 1995 for example). It has however been pointed out in studies of Total Economic Value that option and existence values reflect global as well as local scarcity and that the welfare gained from resource conservation normally accrues at the regional or global level, bypassing the local users (see Adger et al., 1995, and Kumari, 1995, for example).

The crucial aspects of value for local decision-making, and for the differential impacts of global change, are the direct and indirect use benefits. It should be noted that some economic benefits of the mangrove resource will increase in value over time, while others will remain constant or decline. For example, as agricultural development intensifies, the potential economic losses from storm surges increases, so the value of the coastal protection function of the mangroves will rise accordingly. Exogenous environmental change associated with global climate change may increase the frequency and intensity of storm surges, and hence the value of this function of the mangroves will rise.

Recent attempts to value coastal wetlands have concentrated on valuing specific contribution of various wetland systems.Some economics studies have valued the benefits of temperate wetlands, but to date, little analysis of tropical wetland benefit has been undertaken (Turner et al.1991,92). The value of Florida coastal marshes for marine fishery production, relating wetland area to crab production, in this case the Florida blue crab fisheries was estimated by Lynne et al. (1981) and the value of Louisiana wetlands for hurricane wind damage protection was estimated by Gosselink et al. (1974).

E.B. Barbier (1993) gave a basic methodology for assessing and valuing the economic benefit of tropical wetlands applied in Central America in distinguishing between direct-values, indirect-values and non-use values (typology of benefits of conservation), particularly a case study of a mangrove system in Indonesia for woodchip production. The value of wetlands for residential development and the value of Virginia coastal wetlands for oyster production. The valuation of oyster production is directly related to wetland area. The same methodology can therefore be used to value for crab and shrimp fisheries, but not for offshore fisheries were estimated by Bertelson and Shabman (1979).

S. Farber and R. Costanza (1987) stated that the harvest of the shrimp and crabs from the mangroves is directly related to the area of the mangroves by relating the marginal value to the area in a regression equation. The paper derives a linear relationship between area of wetland and offshore fish catch based on a cross section of wetlands. Ruitenbeek, H.J. (1994) gave a full valuation excise, and show the trade-off between protection and logging mangrove forest for timber in indicating that harvesting is much lower when the 'linkages' between the reduction in mangrove area and the functions (such as off-shore fishery) are taken into account. The benefits of the mangroves remaining in existence include the shrimp fishery and biodiversity (Batie and wilson, 1978).

Natural wetlands perform many important functions for humankind-storm prevention, flood and water flow control, and wood products. When properly measured, the TEV may exceed the economic gain of converting the area to an alternative use. It is difficult to assemble estimates of values of these various contributions for the same wetlands system. Because of this difficulty, little is known about the ordes of magnitude of value for one entire wetlands system. Economic analysis of resource use can be undertaken in order to assess the magnitude of benefits to local users of the resource. Some values of the goods and services can be assessed by observation of existing markets, but some of the functions and services of mangroves are indirect, or functional, benefits (see Pearce and Turner, 1990, for example) is still difficult to estimate. The purpose of this paper is to obtain a consistent set of estimates of the value of one wetland system, a biosphere reserve for as many independent contributions as existing methodology and data currently allow.

FRAMEWORK FOR ANALYSIS AT THE STUDY SITE

The valuation of the mangroves is divided into two components: direct use values and indirect or non-use values. For each component, potential counts of benefits as well as their linkage, where the STELLA 2 is basically used for potentially linking components and estimating afterward. For direct use values, all marketable estimations is laid out as thinning, branching, nipa thatching and propagule supply contributing to local benefit 1 though the cost benefit analysis 1. The local benefit 2 includes on site and off site catches, aquaculture at household levels and trade activities of the aquatic products. Clearly, these benefits is linked to other bio-ecological processes such as recovery and resilience of mangroves through biomass and productivity, and nursery and refuse habitats on which the fisheries depend. All linkages should be estimated by ECOPATH application.

For indirect or non use values, the regional benefit that includes air clean, ground water and domestic water supply, and shoreline protection is linked with ecological functions such as gas regulation, recycling of water and erosion and flood control. The global benefits of biodiversity or existence values includes eco-tourism and knowledge benefits. It is a part of eco-tourism may be monetary estimated but the rest and their cultural and information values should be included in the existence values.

Obviously, the plan would be useful to have a analytical framework for study and contribution to management and decision making processes, but we can not evaluate all components presented in the diagram, so a partly estimation is our expectation for doing the project. The linkage should be done for next phases.

PARTICIPATORY ECONOMIC EVALUATION

Context: In the transitional period from the planned to the market economy, the Government of Vietnam has encouraged re-allocation of barren land for smallholder forest resource development. Under the framework of the recent 5-Million-ha Reforestation Program supported by diverse donor agencies, the German-based Kreditanstalt fuer Wiederaufbau (KfW) is co-financing several forestry projects to support smallholder establishment of production forests with protective functions on barren land. Projects are executed by the working group of two consulting firms (GFA Terra systems and GWB).

Forest ecosystem services in the project area are complex in nature and highly site-specific. This makes it difficult to generalize benefits across sites. Also local decision makers and villagers are sceptical about results of economic analysis that is conducted in a style which is not transparent enough for people to understand. People simply do not trust general figures. For instance, they distrust the idea that the environmental value of their forests could be so great in monetary terms. The limitations of conventional valuation suggest that it might be a good thing to encourage active involvement of villagers in the economic valuation of their forests. However, this is a new field in participatory approaches in land use development. To date little practical experience in this field exists in developing countries.

Participatory Economic Valuation (PEV) procedure: The purpose of PEV by the project is to assist forest smallholders and local leaders to improve forest resource development decisions. The basic assumption of PEV is that the best currently available information on the costs and benefits of different land uses is held by local decision-makers, technicians and land users.

The advantages of PEV-compared to conventional valuation approaches-are: (i) PEV fully utilizes the data and knowledge of local decision-makers and villagers; (ii) PEV maximizes the involvement and feedback of these groups; (iii) PEV raises awareness about the importance of forest valuation and knowledge of how to do it. To enable local people to fully participate in valuation, PEV must be conducted in a style which (i) is simple and transparent enough to be understood; (ii) is based on data perceived as accurate; and (iii) uses assumptions perceived to be reasonable.

This can be achieved by building on local knowledge (e.g. of perceived physical environmental impacts in terms of change in yield of rice production by villagers) and economic data and assumptions, combined with field checks with villagers. The core focus of PEV is village valuation by forest extensionists with a group of villagers and commune (township) representatives. All steps in the calculation and results must be visualized and reconfirmed by the participants in the village valuation. It is important to include women in the valuation exercise to ensure that their perspective is not missed. Uncertainties about assumptions made in calculation are addressed by calculating two scenarios (high and low value) for each forest model.

PEV tools used in the PEV exercise included the 3-D model (a three dimensional image of the village), topographical maps, land use maps, ruler, pens, blackboard, and large sheets of paper (A 0).

Forests provide environmental benefits from the village to global scale. The scope of assessment was limited to forest environmental services (FES) for which future local payment is not unlikely. For that reason environmental forest services from which the global community will benefit, but payment is currently not feasible (like carbon sequestration), have been excluded in valuation.

The PEV procedure consists of a minimum of 7 steps, beginning with preparation of background data like collection of secondary yield and market data (step 1), followed by briefing villagers about the purpose, scope and expected results of PEV and collection of basic village data including forest resources (step 2). Then basic village data are collected (step 3). To get a first overview of what type of forest values the villagers perceive, the major direct benefits, forest environmental services, optional values, and non-use values in the village and beyond are identified (step 4). It is important to discuss whether men and women have different perceptions of forest values. To reconfirm the identified forest values, sites are visited and the physical impact area is calculated (step 5). Then, the direct benefits and costs of predominant forest models in the village are calculated, subsequently also environmental services, optional values and non-use values are valued (step 6). Finally the results of the PEV exercise are discussed at a village meeting and recommendations for further actions are made by the villagers (step 7).

PEV field exercise: In December 2001 the PEV concept elaborated by the author during a KfW short term assignment was refined with key project staff and then tested with project extensionists in seven villages in the Provinces of Bac Giang and Lang Son in North Vietnam. For testing the PEV approach only villages were selected where a significant area of the forest canopy was already closed. Results of Exercise: Forest values identified were direct benefits (timber and non-timber products like star anise and resin), environmental benefits (protection of irrigation facilities, improvement of farm land and fish productivity), optional values (forest land reserved for future land use mainly for tree crop development). "Spirit forests" represent the only forest value independent of their actual current and future use (non-use value). Reducing soil erosion and settling sediments in small water reservoirs are the most visible impacts of afforestation of barren slopes. The rate of exchange used was 1 EUR equivalent to 13,000 VND.

Results of calculation of forest environmental services (FES) included: o Enhanced paddy productivity (15,600 to 600,000 VND/ha/year or 1.2 to 46 EUR/ha/year). The highest value is equivalent to the farm gate price of a big pig or the market value of paddy (unshelled rice) from 0.1 ha. o Annual reduced costs for extraction of sedimentation from micro-irrigation facilities (225,000 VND/ha or 17.3 EUR/ha); and Increased annual fish productivity in small village ponds (41,900 VND/ha or 3.2 EUR/ha). This is equivalent to the farm gate price of 4 kg fish.

Optional values for forest land reserved for tree crop cultivation ranged between 8.0 and 94.5 million VND/ha (615 and 7,270 EUR/ha). "Spirit Forests" (non-use value) can have a high value in terms of biodiversity, but villagers did not dare to give a monetary value. The undiscounted Total Economic Value (TEV) per village household in one sample village was between 23.2 and 39.9 million VND/household (1,785 to 3,070 EUR/household) and the discounted TEV between 5.3 and 11.2 million VND/household (410 and 860 EUR/household).

For the high value scenario the undiscounted FEV was around 6% and optional value 1%. The FES for one ha of pine plantation discounted was 14% and FES undiscounted equivalent to 6%. One example of Willingness to Pay (WTP) for disproportionately perceived private environmental forest services (i.e. increased fish pond productivity due to improved water supply and quality from forests) was 10% of the incremental income due to the FES.

Perception of environmental forest benefits was highest where villagers had experienced a period with natural forests, followed by deforestation and unsustainable upland land use including subsequent reforestation. Especially elder villagers highly appreciated the environmental benefits of forests. From the viewpoint of villagers, the distinction between production and protection forests as defined by Vietnamese Law does not make sense in a situation where a production forest in a village also has a significant protective benefit (e.g. when a forested micro-watershed directly protects and improves the productivity of a significant area of paddy land).

Lessons Learned: The exercise in Vietnam reconfirmed that the best realistic economic data for existing forest models can be gained from technicians and farmers, especially those who have developed local forest practices.

The tested bottom-up approach of valuation also proved to be a good medium of communication to identify the production forests in the village for which clear cutting should be avoided at all costs in order to maintain the watershed benefits. In this case PEV provides economic arguments for conversion of unstable pure plantation to continuous forest cover (CFC).

How to deal with time preference, i.e. addressing the fact that people attach greater value to receiving money immediately than they do to obtaining the same amount in the distant future, is still a challenge in the task of further refining participatory resource valuation.

THE PREHISTORY OF COMMUNITY FORESTRY IN INDIA

"Liberty and forest laws are incompatible," remarked an English country vicar, speaking on behalf of villagers shut out of woodland reserved for the exclusive use of the king, in 1720. The history of state forestry is indeed a history of social conflict.

In monarchies and in democracies, in metropolitan Europe as well as in colonial South Asia, the state management of forests has met bitter and continuous opposition. On the one side are the professional foresters who believe that timber production can be ensured only through the exclusion of humans and their animals from wooded areas; on the other, the peasants, pastoralists, charcoal ironmakers, basketweavers, and other such groups for whom access to forests and forest resources is crucial to economic survival. Environmentalists have added to the criticisms of these latter groups, charging foresters with simplifying complex ecosytems in the direction of commercially valuable but biologically impoverished monocultures.

These contending parties have battled for more than two hundred years. In continental Europe, the eighteenth and nineteenth centuries were peppered with social protest movements against the state management of forests. These protests inspired, among other things, Karl Marx's first political writings and a memorable novel by Honore de Balzac capturing peasant hostility to forest officials. When the European model of strict state control over forests was exported to the colonies, the disaffected peasants and tribals responded with arson and violence. Movements over forest rights were a recurring phenomena in colonies ruled by the British, the Dutch, and the French. The conflicts persisted when the post-colonial governments of countries such as Malaysia and Indonesia followed the authoritarian model of forest management inherited from the colonizer.

In recent decades, however, the global discourse on forestry has moved towards a more accommodationist perspective. Foresters and peasant protesters now seem to talk to, rather than talk past, each other. A willingness to listen to and at least partially incorporate the other point of view has replaced the rigid and uncompromising attitude of the past. Within the forestry profession itself, skeptics doubt the contemporary relevance of the custodial and policing approaches previously followed. A system of natural resource management crafted in absolutist and colonialist times clearly needed to be seriously modified or even overthrown.4 Social activists and community leaders have, meanwhile, moved from demanding a total state withdrawal from forest areas to asking governments to more seriously and sympathetically consider the rights of forest-dependent communities.

With this move from conflict to collaboration have come shifts in the language of forestry itself. Terms such as "scientific forestry" and "rational land management," euphemisms for state control and commercial timber production, are being rapidly replaced by sweet-sounding phrases such as "community management," "participatory development," and "joint forest management." While these terms have come into vogue in the last two decades, they have, in fact, a very long geneaology. From the beginnings of state forestry, there have been serious attempts to democratize the regimes of resource management. Both dissidents within the bureaucracy as well as intellectual activists outside it tried hard to make the state respond more sensitively to the just claims of local communities. The ongoing programs of joint forest management in India can draw legitimacy and sustenance from a struggle that is at least a century old.

THE LAW AND THE PROTESTS

The crucial watershed in the history of Indian forestry is undoubtedly the building of the railway network. In a famous minute of 1853, the governor general of India, Lord Dalhousie, wrote of how railway construction was both the means for creating a market for British goods and the outlet for British capital seeking profitable avenues for investment. Thus between 1853 and 1910 more than eighty thousand kilometers of track were laid in the subcontinent.6 The early years of railway expansion witnessed a savage assault on the forests of India. Great chunks of forest were destroyed to meet the demand for railway sleepers (over a million of which were required annually). The sal forests of Garhwal and Kumaun, for example, were "felled in even to desolation." "Thousands of trees were felled which were never removed, nor was their removal possible."

This depradation brought home most forcefully the fact that India's forests were not inexhaustible. At this time the British were unquestionably the world leaders in deforestation, having burnt or felled hundreds of thousands of acres of woodland in Australia, southern Africa, northeastern United States, Burma, and India. Knowing little of methods of sustained-yield forestry, they called in the Germans, who did. Thus in 1864 they established the Indian Forest Department, which for the first twenty-five years of its existence was serenely guided by three German inspectors general of forest-Dietrich Brandis, Wilhelm Schlich, and Bertold von Ribbentrop.

For its effective functioning, the new department required a progressive curtailment of the previously untrammeled rights of use exercised by rural communities all over South Asia. An act was hurriedly drafted to establish the claims of the state to the forestland it immediately required, subject to the provision that existing rights not be abridged. This act was "infinitely milder and less stringent than that which is in force in most European countries." The search commenced for a more stringent and inclusive piece of legislation. In 1869, the Government of India circulated to the provinces a new draft act, which sought to strengthen the state's control over forest areas through the regulation and in some cases extinction of customary rights.

The new legislation was based on the assumption that all land not actually under cultivation belonged to the state. Of course, it was not easy to wish away the access to forests exercised in centuries past by peasants and other rural groups. The colonial state, however, argued that such use, however widespread and enduring, had been exercised only at the mercy of the monarch. Unless it had been expressly recorded in writing, customary use was deemed to be a "privilege," not a "right." And since the British government was the successor to Indian rulers, the ownership of forests and waste was now vested in it. "The right of conquest is the strongest of all rights," emphatically remarked one forest official. "It is a right against which there is no appeal."

There were, however, some notable dissenting voices within the colonial government. Sent the draft forest bill by the Government of India, the Madras Government in turn invited responses from various officers. The views of Narain Row, Deputy

Collector of Nellore, are representative. The proposed legislation, he said, had no historical precedent, for "there were originally no Government forests in this country. Forests have always been of natural growth here; and so they have been enjoyed by the people." Another Deputy Collector, Venkatachellum Puntulu, of Bellary, argued that the burden of the new legislation would fall most heavily on the poor. While large landlords would find it relatively easy to deny the state any claim over their forest property, unlettered peasants would not be able to prove rights of ownership, even though they traditionally used forests as common property.

Criticizing the detailed rules prohibiting the collection of different kinds of forest produce, Puntulu penetratingly remarked that "the provisions of this bill infringe the rights of poor people who live by daily labor (cutting wood, catching fish and eggs of birds) and whose feelings cannot be known to those whose opinions will be required on this bill and who cannot assert their claims, like [the] influential class, who can assert their claims in all ways open to them and spread agitation in the newspapers."

After several such responses came in, the Madras Board of Revenue told the Government of India that the claim of the state to uncultivated forests and wastes was virtually nonexistent:

There is scarcely a forest in the whole of the Presidency of Madras which is not within the limits of some village and there is not one in which so far as the Board can ascertain, the state asserted any rights of property unless royalties in teak, sandalwood, cardamoms and the like can be considered as such, until very recently. All of them, without exception are subject to tribal or communal rights which have existed from time immemorial and which are as difficult to define as they are necessary to the rural population.... [In Madras] the forests are, and always have been common property, no restriction except that of taxes, like the Moturpha [tax on tools] and Pulari [grazing tax] was ever imposed on the people till the Forest Department was created, and such taxes no more indicate that the forest belongs to the state than the collection of assessment shows that the private holdings in Malabar, Canara and the Ryotwari districts belong to it.

The Madras Government advanced three basic reasons for rejecting the bill drafted by the Government of India:

First, because its principles, scope and purpose are inconsistent with the existing facts of forest property and its history. Second, because, even if the Bill were consistent with facts, its provisions are too arbitrary, setting the laws of property at open defiance, and leaving the determination of forest rights to a Department which, in this Presidency at all events, has always shown itself eager to destroy all forest rights but those of Government. Third, because a Forest Bill, which aims at the regulation of local usages ought to be framed, discussed and passed by the local legislature.14

The objections were disregarded, and in 1878, the new bill passed. The act divided the forests of the subcontinent into three broad classes. State or reserved forests were to be carefully chosen, in large and compact areas that could lend themselves to commercial exploitation.

The constitution of these reserves was to be preceded by a legal settlement that either extinguished customary rights of user, transferred them as "privileges" to be exercised elsewhere, or, in exceptional cases, allowed their limited exercise. In the second class, of "protected" forests, rights and privileges were recorded but not settled. However, all valuable tree species were to be declared as "reserved" by the state, while the Forest Department had the power to prohibit grazing and other ostensibly damaging practices.

The Forest Act also provided for a third class of forests-village forests. But as these lands had first to be constituted as reserved forests, the procedure aroused suspicion among the villagers, and this chapter remained a "dead letter."15 Meanwhile, the area of forests under strict state control steadily expanded. In 1878, there were 14,000 square miles of state forest. By 1890, this had increased to 76,000 square miles, three-fourths of which were reserved forests. Ten years later, there were 81,400 square miles of reserved forests and 8,300 of protected forests. Given increasing demand for wood products, the state sought to establish firmer control over forests, both by expanding the area taken over under the Forest Act and by converting protected forests to reserved forest. The Indian Forest Act of 1878 was a comprehensive piece of legislation that came to serve as a model for other British colonies.16 Within India, it allowed the state to expand the commercial exploitation of the forest while putting curbs on local use for subsistence. This denial of village forest rights provoked countrywide protest.

The history of colonial rule is punctuated by major rebellions against colonial forestry-in Chotanagpur in 1893, in Bastar in 1910, in Gudem-Rampa in 1879-80 and again in 1922-23, in Midnapur in 1920, and in Adilabad in 1940. These rebellions sometimes extended over several hundred square miles of territory, involved thousands of villagers, and had to be put down by armed force. Even where discontent did not manifest itself in open rebellion, it was expressed through arson, noncompliance, and breaches of the forest law.

The participants in these protests were unlettered peasants and tribals, and we know far more of their deeds than their words. Nonetheless, their voices do figure here and there in the archives of the state, sometimes mediated by the language of the officials reporting them. Thus in the 1880s, when the government of the Bombay Presidency was aggressively demarcating the rich teak forests of the Dang district, preparatory to their constitution as state reserves, a Bhil tribal chief sent in a petition stating that "we do not wish to let the Dang jungle [be] demarcated, for thereby we shall lose our rights and we and our poor rayat [cultivators] shall always be under the control of the Forest Department and the Department will always oppress us."

Around the same time, the colonial state was attempting to take over the deodar (cedar) forests of the upper Jamuna valley. These trees had suddenly become market-worthy, to service the then expanding railway network. But as a peasant bitterly observed, "the forests have belonged to us from time immemorial, our ancestors planted them and have protected them; now that they have become of value, government steps in and robs us of them."18 Or consider, finally, these remarks of an administrator in the Bastar district of central India, on the determination of his tribal subjects to continue practicing swidden cultivation in what was now "government" forests: "On the road from Tetam to Katekalyan I found general dissatisfaction at the restriction of penda [swidden] cultivation. I was unable to convince them of its evils [sic]. Podiyami Bandi Peda of Tumakpal has to get his son married and for this purpose he wants to cultivate penda in the prohibited area. I told him he should not do it. He replied plainly that he would cultivate it and go to jail as he had to get his son married."

In 1871, the Madras Government predicted that the new act, if passed into law, would "place in antagonism to Government every class whose support is desired and essential to the object in view, from the Zamindar [landlord] to the Hill Toda or Korombar." This was an astonishingly accurate prediction, for the Forest Department was unquestionably the most unpopular arm of the British Raj. The story of the numerous popular movements against state forestry, so long neglected by historians, is now attracting an array of chroniclers. The critics were principally of two kinds. On the one side were scholars and politicians with a deep knowledge of rural conditions, and who sometimes formed part of popular movements themselves.

Their criticisms of state forestry thus drew richly upon the feelings and grievances of the people most affected by it. On the other side were the rare (but, for that reason, significant) dissidents within the colonial bureaucracy, who opposed the centralizing thrust of government forest policy. The first set were outsiders so far as the apparatus of rule was concerned, but insiders with respect to popular opinion and popular consciousness. The second set were by virtue of race and status outsiders to Indian society, but insiders with regard to the policy of the state and the functioning of government.

PRECOCIOUS PROPHETS

In 1878, the Poona Sarvajanik Sabha, a vastly respected nationalist organization in western India, bitterly opposed the new Forest Act. Despite its middle-class origins, the Sabha had consistenly fought for the rights of the cultivator, urging that the colonial government lessen its burden of taxation on the peasantry.21 Now, in the context of the Forest Act debate, it pointed out that state usurpation was grossly violative of customary rights over forests, for both "private grantees and village and tribal communities" had "cherished and maintained these rights with the same tenacity with which private property in land is maintained elsewhere." The Sarvajanik Sabha did not, however, merely oppose the proposed Forest Act for its excessive emphasis on state control; it offered a more constructive and creative alternative.

Thus the Sabha argued that the better maintenance of forest cover could more easily be brought about by taking the Indian villager into confidence of the Indian Government. If the villagers were rewarded and commended for conserving their patches of forestlands, or for making plantations on the same, instead of ejecting them from the forestland that they possess, or in which they are interested, emulation might be evoked between neighboring villages. Thus more effective conservation and development of forests in India might be secured, and when the villagers have their own patches of forest to attend to, government forests might not be molested. Thus the interests of the villagers as well as the government can be secured without causing any unnecessary irritation in the minds of the masses of the Indian population.

The Sabha was advocating a far more democratic structure of forest management than that envisaged by the colonial government. Indeed, it was proposing the institution of a Vrikshamitra (Friends of the Trees) Award, one hundred and ten years before the Indian Government's Ministry of Environment and Forests conceived and named such a scheme, for rewarding individuals and communities who had successfully protected or replenished forest areas.

Three years after the 1878 act was passed, the impact of state forestry on rural communities was foregrounded by the social reformer Jotirau Phule. Phule himself was a gardener by caste, and in general exceptionally alert to the problems of the agricultural classes. The following is his description of the impact of the Forest Department on the livelihood of farmers and pastoralists in the

Deccan countryside: In the olden days small landholders who could not subsist on cultivation alone used to eat wild fruits like figs and jamun and sell the leaves and flowers of the flame of the forest and the mahua tree [both common trees of the Indian forest].

They could also depend on the village ground to maintain one or two cows and two or four goats, thereby living happily in their own ancestral villages. However, the cunning European employees of our motherly government have used their foreign brains to erect a great superstructure called the forest department. With all the hills and undulating areas as also the fallow lands and grazing grounds brought under the control of the forest department, the livestock of the poor farmers do not even have place to breathe anywhere on the surface of the earth.

These remarks drew attention to the dependence of the agriculturist on the produce of forests and other common lands. This dependence was even more acute in the tribal regions of middle India, where communities of hunter-gatherers, swidden agriculturists, and charcoal iron makers were likewise at the receiving end of the new forest laws. These peoples found an eloquent spokesman in Verrier Elwin (1902-1964), a brilliant Oxford scholar and renegade priest who became the foremost interpreter of adivasi (tribal) culture in India. Elwin was a pioneer of ecological anthropology, whose many works vividly showcased the intimate relationship between the forest world and the life of the adivasi. All tribals, he argued, had a deep knowledge of wild plants and animals; some could even read the great volume of Nature like an "open book." Swidden agriculturists, for whom forest and farm shaded imperceptibly into each other, had an especial bond with the natural world. They liked to think of themselves as children of Dharti Mata, Mother Earth, fed and loved by her.

Elwin's ethnographies are peppered with references to the adivasi's love for the forest. Tragically, the forest and game laws introduced by the British had made them interlopers in their own land. He quotes a member of the tribal group Gond, whose idea of heaven was "miles and miles of forest without any forest guards." As the anthropologist himself wrote in 1941:

The reservation of forests was a very serious blow to the tribesman. He was forbidden to practice his traditional methods of cultivation. He was ordered to remain in one village and not to wander from place to place. When he had cattle he was kept in a state of continual anxiety for fear they should stray over the boundary and render him liable to heavy fines. If he was a Forest Villager he became liable at any moment to be called to work for the Forest Department. If he lived elsewhere he was forced to obtain a license for almost every kind of forest produce. At every turn the Forest Laws cut across his life, limiting, frustrating, destroying his self confidence.

During the year 1933-4 there were 27,000 forest offences registered in the Central Provinces and Berar and probably ten times as many unwhipped of justice. It is obvious that so great a number of offences would not occur unless the forest regulations ran counter to the fundamental needs of the tribesmen. A Forest Officer once said to me: "Our laws are of such a kind that every villager breaks one forest law every day of his life."

Elwin's writings were addressed equally to the colonial state and to the Congress nationalists, who in the 1940s were very much a government-in-waiting. The Congress, however, had not been especially sensitive to the rights of the tribals. But as Elwin reminded them, "the aboriginals are the real swadeshi [indigenous] products of India, in whose presence everything is foreign. They are the ancient people with moral claims and rights thousands of years old. They were here first: they should come first in our regard." He was deeply distressed when a Congress report on tribals followed the British authorities in asking for a ban on shifting cultivation.

Now Elwin's work had shown that, contrary to modernist prejudice, swidden as practiced by the Baiga, the Juang, and other tribes was an ecologically viable system of cultivation. When the nationalists recommended the ban, he wrote angrily that "the forests belong to the aboriginal. I should have thought that anyone who was a Nationalist would at least advocate swaraj [freedom] for the aboriginal!"

The significance of the forest in tribal life is a running theme in Elwin's work. Noting that a majority of tribal rebellions had centered around land and forests, he pleaded for the greater involvement of tribals in forest management in free India. Even if adivasis had no longer any legal rights of ownership, they had considerable moral rights. And as tribals were as much part of the national treasure as forests themselves, there should be an amicable adjustment between forest management and tribal needs. Even where commercial forest operations became necessary, he said, these should be undertaken by tribal cooperatives and not by powerful private contractors.

After independence Verrier Elwin became the first foreigner to be granted citizenship of free India. In 1954, he was appointed Adviser on Tribal Affairs to the Government of India (with special reference to the North-east Frontier Agency). He was also to serve on more than one high-level, all-India committee on tribal policy. From his first official appointment until his premature death in 1964, Elwin repeatedly urged a reconsideration of forest policy, such that it might, at last, come to more properly serve tribal needs. In this he had little success for forest management became, if anything, more commercially oriented in independent India.32 Towards the end of his life, the anthropologist wrote with some bitterness of how the victims of government policy were being unfairly blamed for the destruction of forests:

There is constant propaganda that the tribal people are destroying the forest. When this was put to some of the villagers, they countered the complaint by asking how they could destroy the forest. They owned no trucks; they hardly had even a bullock-cart; the utmost that they could carry away was a headload of produce for sale to maintain their families and that too against a license. The utmost that they wanted was wood to keep them warm in the winter months, to reconstruct or repair their huts and carry on their little cottage industries. Their fuel-needs for cooking, they said, were not much, for they had not much to cook. Having explained their own position they invariably turned to the amount of [forest] destruction that was taking place all around them. They asked how the exzamindars [landlords], in violation of their agreements and the forest rules and laws, devastated vast tracts of forest land right in front of officials. They also related how the contractors stray outside the contracted coupes, carry loads in trucks in excess of their authorised capacity and otherwise exploit both the forests and the tribal people.

There is a feeling among the tribals that all the arguments in favor of preservation and development of forests are intended to refuse them their demands. They argue that when it is a question of industry, township, development work or projects of rehabilitation, all these plausible arguments are forgotten and vast tracts are placed at the disposal of outsiders who mercilessly destroy the forest wealth with or without necessity.

From a great Englishman who devoted his life to the service of the Indian poor, we move on to a great Englishwoman who did likewise-Madeleine Slade, the daughter of an admiral who came from England to join Gandhi in his Sabarmati Ashram in 1926. Gandhi adopted her as his own daughter and gave her the name Mira Behn. She played a prominent role in the anti-colonial struggle and was jailed several times.

In 1945, Mira Behn set up a Kisan (peasant) Ashram near the holy town of Hardwar, and two years later moved up the Ganges beyond Rishikesh, where the river descends into the plains. In 1952, she shifted her base again, to the Bhilangna valley in the interior Himalaya. Here she stayed still 1959 when ill health and possibly dissatisfaction with the policies of independent India made her migrate to Austria.

The peasants of the central Himalaya are, of course, as dependent on forest produce as the tribals of the Indian heartland with whom Elwin long worked. Here, one unfortunate consequence of state forest management was the gradual replacement of banj oak (quercus incana), a tree much prized by villagers as a source of fuel, fodder, and leaf manure, by chil (or chir) pine (pinus roxburghii), a species more valued commercially as a source of timber and resin. This transition had serious ecological implications, for the thick undergrowth characteristic of banj forests absorbed a high proportion of the rainwaters of the fierce Himalayan monsoon. This water then slowly percolated downhill. Below the oak forests were thus found "beautiful sweet and cool springs," the main source of drinking water for the hill villagers. By contrast, the floor of pine forests was covered thinly by needles, and had much less absorptive capacity. In hillsides dominated by chil, the rain rushed down the slopes, carrying away soil, debris, and rock, contributing thereby to floods.

Why were the banj forests disappearing in the Himalaya? Mira Behn's own explanation revealed a sharp awareness of the sociology of forest management in the hills. "It is not merely that the Forest Department spreads the Chil pine," she said, "but largely because the Department does not seriously organize and control the lopping of the Banj trees for cattle fodder, and... is glad enough from the financial point of view to see the Banj dying out and the chil pine taking its place. When the Banj trees grow weak and scraggy from overlopping, the chil pine gets a footing in the forest, and once it grows up and starts casting its pine needles on the ground, all other trees die out."

Mira Behn continued: "It is no good putting all the blame on the villagers....

The villagers themselves realize fully the immense importance of these Banj forests, without which their cattle would starve to death, the springs would dry up, and flood waters from the upper mountain slopes would devastate their precious terraced fields in the valleys. Indeed all these misfortunes are already making their appearance on a wide scale. Yet each individual villager cannot resist lopping the Banj trees in the unprotected Government forests. 'If I do not lop the trees someone else will, so why not lop them, and lop them as much as possible before the next comer.'"

Although Mira Behn does not explicitly make the point, it seemed that this shortsighted behavior of the hill peasant was related to the loss of community control, such that individual peasants no longer had a long-term stake in the maintenance of forest cover. This was a tendency aggravated by the commercial orientation of the Forest Department. Could anything be done to restore banj to its rightful place, and thus revive Himalayan economy and ecology? Mira Behn writes: The problem is not without solution, for if trees are lopped methodically, they can still give a large quantity of fodder, and yet not become weak and scraggy. At the same time, if the intruding Chil pines are pushed back to their correct altitude (i.e. between 3,000 and 5,000 feet), and the Banj forests are resuscitated, the burden on the present trees will, year by year, decrease, and precious fodder for the cattle will actually become more plentiful.

But all this means winning the trust and co-operation of the villagers, for the Forest Department, by itself, cannot save the situation. Nor can it easily win the villagers' trust, because the relations between the Department and the peasantry are very strained, practically amounting to open warfare in Chil pine areas. Therefore, in order to awaken confidence in the people, some non-official influence is necessary. With the aid of local constructive workers, it should become possible to organize village committees and village guards to function along with the Forest Department field staff which should be increased, and also given special training in a new outlook towards the peasantry.

In this way it should be feasible to carry out a wellbalanced long term project for controlled lopping and gradual return of the Banj forests to their rightful place, by systematic removal of Chil pines above 5,500 feet altitude to be followed by protection of the young Banj growth. The Banj forests are the very centres of nature's economic cycle on the southern slopes of the Himalayas. To destroy them is to cut out the heart and thus bring death to the whole structure.

Mira Behn sent reports of her findings, with photographs, to Prime Minister Jawaharlal Nehru. He passed them on to the concerned officials, but nothing seems to have come of it; it appears that the Indian Forest Department of the 1950s would not change its ways.

A DEMOCRATIZING FORESTER

The quotes offered in the previous section are all strikingly contemporary. To those who know something of the people behind them, they are also perfectly in character. Phule, Elwin, and Mira Behn, as well as the leaders of the Poona Sarvajanik Sabha, had a deep knowledge of agrarian life. Alert to the inequities in access to natural resources brought about by the new laws, they would vigorously polemicize on behalf of the victims of state forest management. Dietrich Brandis was a prophet of community forestry who came from the unlikeliest of backgrounds. He was a forest officer; in fact, no less than the first inspector general of forests (IGF) in India. In nineteen years as IGF (1864-1883), he laid the foundations of state forestry in India.

A man of great energy, he toured widely in the subcontinent, writing authoritative reports on the direction forest management should take in the different provinces of British India.36 In the realm of silviculture, he formulated the systems of valuation and forest working still widely in use. As a former university don himself (he came to the service of the Raj from the University of Bonn), Brandis started a college for training subordinate staff, arranged for higher officials to be trained on the continent, and helped set up the Forest Research Institute in Dehradun.

The scientific and administrative aspects of Brandis's legacy are not our focus here, but rather his sociology of forest management, his understanding of the social and political contexts within which state forestry had to operate in India.37 Here Brandis's views must be immediately distinguished from almost all other forest officials, Indian or European, before or since. These officials counterpose "scientific" forestry under state auspices to the customary use of forests by rural communities, which they have always held to be erratic, unsystematic, wasteful, and shortsighted. It is thus that the forest officials justify their territorial control of over one-fifth of India's land mass, claiming that they alone possess the technical skills and administrative competence to manage woodland.

To be sure, Dietrich Brandis shared this creedal faith in the scientific status of sustained-yield forestry. He also believed that the state had a central role to play in forest management. But what he certainly did not share was his colleagues' skepticism of the knowledge base of rural communities. For example, Brandis wrote appreciatively of the widespread network of sacred groves in the subcontinent. These he termed, on different occasions, the "traditional system of forest preservation" and examples of "indigenous Indian forestry." In his tours he found sacred woodlands "most carefully protected" in many districts-from the Devara Kadus of Coorg in the south to the deodar temple groves in the Himalaya.

At the other end of the social spectrum, Brandis also wrote appreciatively of forest reserves managed by Indian chiefs. He was particularly impressed by the Rajput princes of Rajasthan, whose hunting preserves provided game for the nobility as well as a permanent supply of fodder and small timber for the peasantry. The British stereotype of the Indian Maharaja was of a feckless and dissolute ruler, but as Brandis pointed out, in strenuously preserving brushwood in an arid climate the Rajputs had "set a good example, which the forest officers of the British government would do well to emulate."

In Brandis's larger vision for Indian forestry, a network of state reserves would run parallel to a network of village forests. The Forest Department would take over commercially valuable and strategically important forests, while simultaneously encouraging peasants to collectively manage areas left out of these reserves. Through a series of reports and memoranda written over a decade, the IGF tried to persuade the colonial government that a strong system of village forests was vital to the longterm success of state forestry itself.

The first such report was written in 1868, and pertained to the southern province of Mysore. This was a closely argued document suggesting the creation of village forests throughout Mysore, managed on a rotational cropping system, with freshly cut areas closed to fire and grazing. Ideally, each hamlet would have its own forest, but in many cases it might become necessary to constitute a block to be used by a group of villages. Such forests would provide the following items free of cost-firewood for home consumption and for sale by "poor people with headloads"; wood for agricultural implements and the making and repairing of carts; wood, bamboo, and grass for thatching, flooring, and fencing; leaves and branches for manure; and grazing except in areas closed for reproduction. On the payment of a small fee, wood would be made available for houses and for use by artisans.

In Brandis's scheme, these forests would be put under a parallel administrative system, with a village forester for each unit, a forest ranger for all the village forests in each taluk (county), and a head forest ranger for the district as a whole, this man reporting to an assistant conservator of forests. He anticipated that the system would be self-supporting, with any surplus used for local improvements. In this manner, peasants would come to feel an interest "in the maintenance and improvement of their forests." Brandis also hoped that Forest Department control over village forests would give way in due course, with the "leading men" of each village assuming responsibility for management.

Forwarding his report to the Government of India, Brandis noted significantly that it was "the first of a series of measures" which he proposed "to suggest in various Provinces for the better utilization and for the improvement of the extensive wastelands which will not be included in the State Forests": that is, as a prelude to recommending a countrywide system of community forests. Unhappily, the British officials of the Raj lacked Brandis's understanding of the biomass economy of rural India, the vital dependence of agrarian life on the produce of the forests. They also lacked his faith in local knowledge and local initiative.

The opposition to Brandis claimed that his scheme would lead to a loss of state revenues while undermining the powers of district officials. Also invoked was an early version of the "Tragedy of the Commons" argument. For one official, "the village communities of Mysore, without cohesion and often split up into factions by caste, could not be entrusted with the powers, or competent to perform the functions assigned to them in [Dr Brandis's] scheme." Another commented that the scheme would fail "as each man, when the least removed from supervision, would cut whatever he might require for himself without any regard to the interests of his neighbours." The Government of India's final, negative verdict rested on a classic piece of colonial stereotyping. "The prejudices and rivalries of Natives," it said, "might be excited if men of different classes and castes shared in the same forests."

Brandis did not lightly accept this judgment. In a defiant note, he reviewed the case afresh, and made another forceful plea in favor of village forests. He drew pointed attention to the flourishing system of community forests on the continent, where scientific foresters exercised technical supervision over woodland managed for the exclusive benefit of villages and small towns. In Europe, wrote Brandis, "Such Communal Forests are a source of wealth to many towns and villages in Italy, France and Germany; property of this nature maintains a healthy spirit of independence among agricultural communities; it enables them to build roads, churches, school-houses, and to do much for promoting the welfare of the inhabitants; the advantages of encouraging the growth, and insisting on the good management of landed communal property, are manifold, and would be found as important in many parts of India as they have been found in Europe."

Following his failure in Mysore, Brandis resurrected his proposals in the debate leading up to the 1878 Forest Act. Here he urged the administration "to demarcate as state forests as large and compact areas of valuable forests as can be obtained free of forest rights of persons," while leaving the residual area, smaller in extent but more conveniently located for their supply, under the control of village communities. He hoped ultimately for the creation of three great classes of forest property, based on the European experience: state forests, forests of villages and other communities, and private forests. State ownership had to be restricted, argued Brandis, on account of the "small number of experienced and really useful officers" in the colonial forestry service and out of deference to the wishes of the local population. For "the trouble of effecting the forest rights and privileges on limited well-defined areas is temporary and will soon pass away, whereas the annoyance to the inhabitants by the maintenance of restrictions over the whole area of large forest tracts will be permanent, and will increase with the growth of population."

Here was an uncanny anticipation of the widespread popular opposition that has been such a marked feature of the subsequent history of Indian forestry. But Brandis was overruled by more powerful civil servants within the colonial bureucracy, and the 1878 Forest Act was based firmly on the principle of state monopoly. But the German forester was a remarkably persistent man. As he remarked shortly after relinquishing the post of inspector general, systematic forestry in India "was like a plant of foreign origin, and the aim must be to naturalize it."

On the social side, this process of indigenization could be accomplished by encouraging native chiefs, large proprietors, and especially village communities to develop and protect forests for their own use. In the last instance the initiative lay with the government, which, insisted Brandis, stood to gain enormously from a successful system of communal forests. "Not only will these forests yield a permanent supply of wood and fodder to the people without any material expense to the State," he wrote, "but if well managed, they will contribute much towards the healthy development of municipial institutions and of local self-government."

In 1897, well into his retirement in Germany, Brandis returned to the subject of community forests. Long after he had severed all formal contacts with British India, Brandis continued to be deeply concerned that Indian forestry should cease to have "the character of an exotic plant, or a foreign artificially fostered institution." This concern was consistent with his larger democratic vision for forestry in the subcontinent. Thus he suggested that "native" forest officers, as they distinguished themselves, be sent to study the forestry system operating in Germany. Notably, Brandis had in mind their social as well as silvicultural education. As he concluded his essay, Indian foresters, if sent to Germany, "will find that the villages, which own well-managed communal forests, are prosperous, although now and then they complain of the restrictions that a good system of management unavoidably imposes. What Indian forest officers will learn in this respect in Germany will be really useful to them in India."

Perhaps by now Brandis despaired of British officials in India taking seriously his proposals for the constitution of village forests. Hence this indirect approach, wherein Indian forest officers trained on the continent might be able to better see the benefits of community forests. In the event, Indian officials (whether trained in Germany or not) have been, for the most part, hostile to any suggestion that local communities could be encouraged to manage forest areas for their own use. It is, indeed, this territorial monopoly and indifference to the demands of rural communities that have made the Forest Department the object of such relentless criticism in recent years.

The Indian Forest Department has been the subject of sharp attack for its authoritarian style of functioning; and yet, in an interesting paradox, the founder of the department had himself anticipated that a narrow reliance on state control and punitive methods of management would lead to popular disaffection. While terms such as "social forestry," "community forestry," and "joint forest management" have only now come into currency, the principles they embody would have been readily recognized, and indeed warmly commended, by the first head of the Forest Department in India.

THE HIMALAYAN CASE

For all their insight, knowledge, and passion, these precocious advocates of community forestry did not have much impact on state policy. Control and commercialization remained the dominant motifs of state forest policy. The chapter on village forests in the 1878 act remained a dead letter. Government forest policy, in the colonial as well as postcolonial periods, continued to seriously ignore village needs, demands, and interests. The principle of state monopoly has remained paramount, with one very partial exception.

The Kumaun and Garhwal hills of present-day Uttar Pradesh contain the best stands of softwood in the subcontinent. These coniferous species have been highly prized since the early days of colonial forest management. Between 1869 and 1885, for example, some 6.5 million railway sleepers made from deodar (cedrus deodarus, the Indian cedar) were exported from the valley of the Yamuna, in the princely state of Tehri Garhwal. Adjoining Tehri Garhwal to the east was the British-administered Kumaun division, with its rich stands of chir pine. Here forestry operations concentrated simultaneouly on expanding the area under chir (at the expense of oak) and exploiting the tree both for timber and for resin. Between 1910 and 1920, for example, the number of trees tapped for resin increased from 260,000 to 2,135,000.

The pine trees of the Central Himalaya were the only source, within the British Empire, of oleo-resin, an extract with a wide range of commercial and industrial applications. Likewise, the timber of deodar and chir, as well as fir and spruce, constituted a strategically valuable resource for the colonial state, exploited with profit to service the military campaigns of the two world wars.

In the Himalaya, as elsewhere, commercial forestry under state auspices was made possible only through a denial of customary rights of ownership and use. In these hills, forests and grassland were a crucial resource for the agro-pastoral production system.

In fact, the fragmentary evidence available to historians does suggest the existence of a fairly widespread system of common property resource management-with grass reserves walled in and well looked after, oak forests managed by the village community, and sacred groves lovingly protected. Not surprisingly, the government's attempts to seize vast areas under local control and reconstitute them as "reserved forests" evoked opposition. In the early years of state management, a petition from a discontented hillman evocatively recalled a golden age when the villagers had full control over their forest habitat:

In days gone by every necessities of life were in abundance to villagers than to others [and] there were no such government laws and regulations prohibiting the free use of unsurveyed land and forest by them as they have now. The time itself has now become very hard and it has been made still harder by the imposition of different laws, regulations, and taxes on them and by increasing the land revenue. Now the village life has been shadowed by all the miseries and inconveniences of the present day laws and regulations.

They are not allowed to fell down a tree to get fuels from it for their daily use and they cannot cut leaves of trees beyond certain portion of them for fodder to their animals. But the touring officials still view the present situation with an eye of the past and press them to supply good grass for themselves and their [retinue] without thinking of making any payment for these things to them who after spending their time, money and labour, can hardly procure them for their own use.

In short all the privileges of village life, as they were twenty years ago, are nowhere to be found now, still the officials hanker after the system of yore when there were everything in abundance and within the reach of villagers.

When such protests went unheeded, the sentiments underlying them were to manifest themselves in sustained and organized resistance on the part of the Himalayan peasantry. In fact, this region probably witnessed more and more serious social conflict than any other forest region of India.

There were major peasant movements against state forest policies in 1904, 1906, 1916, 1921, 1930, and 1942. These recurrent conflicts, remarked one sensitive official, were a consequence of "the struggle for existence between the villagers and the Forest Department; the former to live, the latter to show a surplus and what the department looks on as efficient forest management."

The most significant forest movement in Kumaun and Garhwal took place in 1921. This took the shape of labor strikes, which crippled the administration, and the widespread burning of pine forests. A total of 395 recorded fires burnt an estimated 246,000 acres of forest. Hundreds of thousands of resin channels were destroyed.

Constituting a direct challenge to the state to relax its control over forest areas, these protests enjoyed enormous popular support, which made it virtually impossible for the administration to detect the people responsible for the fires. The fires were generally directed at areas where the state was at its most vulnerable, for example, compact blocks of chir forest worked for timber and/or resin. Significantly, there is no evidence that the large areas of broad-leaved forest, also controlled by the state, were at all affected. Thus arson was not random but carefully discriminating-it spared those species more useful to the village economy.

In the vanguard of the 1921 movement were soldiers who had fought for the British in the First World War. Kumaun and Garhwal had long supplied hardy and exceptionally brave soldiers for the British Army-indeed, three of the five Victoria Crosses awarded to Indians between 1914 and 1918 went to this region.

These former men in uniform saw the forest regulations as a bitter betrayal of their interests by the white overlord for whom they had so recently risked their lives. Their protests alarmed the colonial state, for apart from being a reservoir of able-bodied men whom it hoped to continue to recruit for the wars it had to fight, the Kumaun hills bordered both Nepal and Tibet-regions not under direct British suzerainty but in which it had strong trading and political interests.

In the wake of the popular protests, a magisterial critique of government forest policy was published by Govind Ballabh Pant. Pant was a rising lawyer from a peasant household in Almora who went on to become one of the foremost of Indian nationalists, after independence taking office successively as chief minister of Uttar Pradesh and home minister of the Government of India.

His 1922 booklet The Forest Problem in Kumaun described the "burial of the immemorial and indefeasible rights of the people of Kumaun," buried, that is, "between the property-grabbing zeal of the revenue officers and the exhortations of experts of the forest department." As he put it, with legal precision, "the policy of the Forest Department can be summed up in two words, namely, encroachment and exploitation."

Several decades of a single-minded commercial forestry had led to a manifest deterioration of the agrarian economy: "Symptoms of decay are unmistakably visible in many a village: buildings are tottering, houses are deserted, population has dwindled and assessed land has gone out of cultivation since the policy of [forest] reservation was initiated.... Cattle have become weakened and emaciated and dairy produce is growing scarce every day: while in former times one could get any amount of milk and other varieties for the mere asking, now occasions are not rare when one cannot obtain it in the villages, for any price for the simple reason that it is not produced there at all."

Pant's analysis was rooted in a deep knowledge of the local context. He took it upon himself to combat the charge, commonly levied against the hill peasant, "of reckless devastation," a charge "sedulously propagated by prejudiced or ignorant persons." As he wrote, The spacious wooded areas extending over the mountain ranges and hill sides bear testimony to the care bestowed by the successive generations of the Kumaonies.

All of them are not of spontaneous growth and specially the finer varieties bespeak his labour and instinct for the plantation and preservation of the forest. A natural system of conservancy was in vogue, almost every hill top is dedicated to some local deity and the trees on or about the spot are regarded with great respect so that nobody dare touch them.

There is also a general impression among the people that everyone cutting a tree should plant another in its place.... Grass and fodder reserves are maintained, and even nap [cultivable] lands are covered with trees, wherever, though in few cases, such land could be spared from the paramount demand of cultivation. Special care is also taken by the villagers to plant and preserve trees on the edges of their fields.

From this analysis, the solution logically offered was to give back to the peasants the woodland that they traditionally regarded as being within their village boundaries.

"If the village areas are restored to the villagers, the causes of conflict and antagonism between the forest policy and the villagers will disappear, and a harmony and identity of interests will take the place of the distrust, and the villager will begin to protect the forests even if such protection involves some sacrifice or physical discomfort." Pant envisaged that these areas would be under the control of the village panchayats, or councils, under whose direction the "natural system of conservancy" would once again come to the fore. As he shrewdly observed, "some restrictions will be there, but these will proceed from within, and will not be imposed from without."

Clearly, Pant drew upon and systematized the knowledge, perceptions, and analysis of the peasant folk of the Kumaun Himalaya. Thus, after the popular protests in 1921, the Government of the United Provinces set up a Kumaun Forest Grievances Committee. This committee toured the hills, examining some five thousand witnesses in all.

Peasant activists submitted dozens of petitions to the committee, on behalf of individual villages. These identified blocks of forest near every village, where peasants would have exclusive rights of fuel and fodder collection, timber for building, wood for ploughs, bamboos for basket making, etc. It was being proposed that villagers should have full rights over these forests, which they would manage through their own panchayat.

Based on the evidence it collected, the committee finally concluded that "any attempt to strictly enforce these [forest] rules would lead to riots and bloodshed."

It thus divided the existing reserved forests into two categories-Class I, which were to be managed not by the forest officials but by the civil administration (in theory more sympathetic to rural needs), and Class II, constituting the commercially valuable wooded areas, which were to remain with the Forest Department. It also recommended that the government consider the constitution of village forests as per the demands of the people of Garhwal and Kumaun.

Bureaucracies move at their own pace, and only in 1930 the rules were passed allowing for the formation of van panchayats, or village forests, in the hill districts of the United Provinces. These allowed for a forest patch to be handed over to a village if it lay within its settlement boundaries, and if more than one-third of its residents had applied for permission to the deputy commissioner (DC). Once the DC gave the go-ahead, then the villagers elected, by voice vote; a council (panch) of five to nine members. This council in turn elected a head (sarpanch) among themselves. The van panchayat was empowered to close the forest for grazing, regulate cutting of branches and collection of fuel, and organize the distribution of forest produce.

It could appoint a watchman, whose salary would be paid by villagers' contributions to the panchayat. The panchayat could levy fines, although if the offender did not pay it had then to go to the civil courts for redressal.

The felling of trees, however, required the permission of the Forest Department. The department also claimed 40 percent of the revenue from any commercial exploitation. Of the rest, 20 percent would go to the zilla parishad (district council), with the balance 40 percent kept with the DC on behalf of the van panchayat, which with that official's written consent could use the funds for roads, schools, and other local improvements.

There are now in excess of 4,000 van panchayats in Kumaun and Garhwal, covering an area of just less than half a million hectares. An official report of 1960 remarked that many of these village forest councils had done "exemplary work in connection with forest protection and development." A more recent survey has concluded that the panchayat forests are often in a better condition than the reserved forests. Of twenty-one panchayats surveyed in three districts, the forest stock in thirteen of these were in good condition, in four in medium condition, in three in poor condition.

The researcher concluded that van panchayats have, by and large, maintained oak forests very well, especially in contrast to the dismal condition of the reserves (except for those reserves distant from habitations). The position in respect of chir forests is not so clear, but these seem to have done about as badly under van panchayat control as in the reserves. Various studies suggest that, overall, panchayat forests seem to be in as good or better condition than the reserves.

The van panchayat system constitutes the only network of village forests mandated by law in all of India.61 The concession was made by a colonial state worried of losing control in a sensitive and strategically important border region, and it was not to be replicated elsewhere. After independence, the van panchayat regime was not extended to the adjoining region of Tehri Garhwal, where Mira Behn worked in the 1950s, and where it might have very well contributed to preserving and enhancing the oak forests. Within Kumaun, too, there is considerable resentment over the curbs placed on the autonomous functioning of van panchayats. Though technically under the control of the villagers, the Forest Department can veto schemes for improvement, while of the revenue generated, 40 percent is swallowed by the state exchequer. Forty percent of the rest is by law granted to the village, but this money too first finds its way into a "consolidated fund" controlled by the DC, to which individual panchayats have then to apply.

There are signs of an emerging movement to do away with these constricting rules, to make the management of the panchayats come fully under the control of the villagers. A chronicler of this discontent, himself quite aware of the long history of forest-related protests in Kumaun and Garhwal, writes that "those who know the history of forest struggles say that... the van panchayat movement will be the biggest such movement in the hills."

TWO CHEERS FOR JOINT FOREST MANAGEMENT

From the inception of state forestry in India, perceptive critics have argued for a democratization of resource control, for a correction of the commercial bias promoted by successive governments, and for a proper participation in management and decision making by local user groups. Arguments first offered in the 1870s, and reiterated in subsequent decades, were revived, or reinvented, in the 1970s by the now-famous Chipko movement. It is no accident that Chipko originated in Garhwal and Kumaun, the part of India that has seen some of the most intensive conflicts between the state and the peasantry over forest resources.

The 1970s were marked by a series of forest movements in different parts of India. These took place in the Himalaya, in the

Western Ghats, and, above all, in the vast tribal belt extending across the heart of peninsular India.

In the Chotanagpur plateau, forest protests formed an integral part of the larger movement for a separate tribal homeland of Jharkhand, carved out from the huge, unwieldy, and predominantly non-tribal state of Bihar. In one much celebrated case, tribals demolished a plantation of teak, a highly prized furniture wood, that was coming up on land previously under the sal tree (Shorea robusta), a species of far greater benefit to the local economy.

Their slogan, "Sal means Jharkhand, sagwan (teak) means Bihar" was a one-sentence critique of the narrow commercial ends of state forestry. Since the 1970s, there has been an ongoing, nationwide debate on forest policy in India, a debate fuelled by the continuing social tension in forest areas and the evidence of massive deforestation provided by satellite imagery.

This debate has passed through three distinct if chronologically overlapping phases. The first phase might be designated the "politics of blame." The activists speaking on behalf of disadvantaged groups have held the forest officials responsible for environmental degradation and popular discontent. The officials, in turn, have insisted that growing human and cattle populations are the prime reason why fully half of the 23 percent of India legally designated as "forest" was without tree cover.

The forestry debate of the 1970s and the 1980s drew, at times, on the heritage of earlier movements and critiques. The peasants of Garhwal and Kumaun, as this writer found out while doing field work there in 1982-83, were acutely conscious of how Chipko itself drew on a long and honorable history of peasant resistance to state forestry.

Tribal activists in Madhya Pradesh and Bihar, meanwhile, were not unfamiliar with the work and message of Verrier Elwin. And in the villages of the Deccan, social workers liked to offer the same quote of Jotirau Phule's reproduced earlier in the paper, as proof that in the agro-pastoral system of that region, proper access to forests and pasture was vital to survival, and that it was the "great superstructure" of the Forest Department that continued to deny herders and farmers this access.

However, perhaps the most direct connection between the past and the present of forest management was effected in the summer of 1982, when the Government of India circulated a new draft forest act. Activists and academics joined hands to demonstrate how the proposed legislation was solidly based upon and, indeed, took further forward the centralizing thrust and punitive orientation of the notorious Indian Forest Act of 1878. After a countrywide campaign, the draft bill was finally dropped by the state.

As tempers cooled and polemic exhausted itself, a second phase, the "politics of negotiation," originated. In villages and state capitals, forest officers and their critics found themselves at the same table, talking and beginning to appreciate, if not fully understand, the other's point of view. Concessions were made by each side, protests suspended by one, and leases of forest produce to industry cancelled by the other.

One product of the growing dialogue between activists and bureaucrats was the approval, by the Indian Parliament in 1988, of a new National Forest Policy. Where the ruling Forest Policy of 1952 had stressed state control and industrial exploitation, the new document instead emphasized the imperatives of ecological stability and peoples' needs.

Then, slowly and hesitatingly, commenced the third phase, "the politics of collaboration." In the state of West Bengal, for example, the Forest Department initiated remedial action on its own, abandoning its traditional custodial approach by inviting peasants to cooperate with it. Thousands of village forest protection committees were constituted, each of which pledged to protect nearby forests in collaboration with the state. Thus previously authoritarian government officials joined with previously suspicious villagers to succesfully regenerate the degraded sal forests of southwestern Bengal.

The success of "Joint Forest Management" or JFM in West Bengal has encouraged scholars, activists, and sympathetic civil servants to demand its replication in other parts of India. Outside its original home, however, the progress of JFM has been slow. Administrative styles and cultures of governance vary widely among the states and regions of India. So do individual orientations, with some forest officials still loathe to relinquish control, while others have been inspired to start village protection committees on their own.

A mapping of the forestry debate in contemporary India would therefore show significant regional variations. Some states are still stuck in the "politics of blame"; others have moved tentatively to the "politics of negotiation." West Bengal and parts of Andhra Pradesh, Madhya Pradesh, and Himachal Pradesh have instituted the "politics of collaboration" through the creation of JFM regimes. In this last scenario there is abundant scope for improvement. As analysts have shown, the JFM model now promoted by the Government of India reflects and sometimes reinforces inequities within rural society.

Gender and caste are two axes of discrimination, with women and low-caste members of the village community not having adequate representation or voice in the decision-making process (this is also true, to a great extent, of the van panchayats in Kumaun.) Likewise, pastoral groups and artisans, who have legitimate claims on forest resources, are sometimes given short shrift. Moreover, the forest officials still claim a monopoly of "scientific expertise," refusing to entertain villagers' own ideas on species choice, spacing, or harvesting techniques.

One serious problem with the JFM model, as currently promoted by the state and donor agencies, is that it allows the constitution of village forest committees only on forestland with less than 40 percent crown cover. This is a deeply constricting rule, which reserves to the state, and the state alone, exclusive rights over the bestclothed lands of India.

Thus forests situated close to hamlets cannot come under JFM regimes if they have more than 40 percent tree cover. Again, the regulations, strictly interpreted, would mean that if local communities were to effectively protect and replenish degraded lands, such that the crown cover was to come to exceed that magic figure of 40 percent, the state could step in and remove the area from JFM-which would be a bizarre outcome indeed.

Nor have changes in policy and orientation been accompanied by concomitant changes in legislation. Thus, the present regime is not flexible enough to allow for spontaneous community-initiated forest regimes to exist along with more orthodox JFM regimes. In some parts of India, the Forest Department is casting a covetous eye on areas well protected by village communities.

Thus in the Uttar Pradesh hills, the old established panchayat forests, managed by villagers, are sought to be brought under the JFM system only so that bureaucrats would have a greater say in their management. A new, carefully thought out Indian Forest Act is called for, which allows both for areas to be managed under state-village partnerships as well as by self-generated, autonomous community regimes.

One can thus envision a fourth (and possibly final) phase for the Indian forestry debate, the "politics of partnership." For collaboration, even where it does exist, takes place on terms set down by the state, through the officials of the Forest Department. We need to move on to a more inclusively democratic structure, where the state listens to and learns from the community, and where the community itself recognizes and deals fairly with the inequities within its own ranks.

The evidence suggests that contemporary advocates of decentralized forestry have had far greater success than their precursors. One reason for this is the altered political context: Brandis, Pant, and company worked under a colonial, authoritarian regime; the partisans of Chipko and similar movements in a democratic system.

The revival of forest protest in the 1970s also coincided with the international environmental debate, which foregrounded the use and abuse of forests worldwide. The work of Indian scholars had, meanwhile, demonstrated with authority that the century-old history of state forestry in India must be reckoned a failure, in both an ecological and social sense. Finally, the problems with government-directed development programs in much of the Third World had led to an increasing interest in nongovernmental forms of management and control.

These calls for forest reform from the outside were complemented by pressures from change from within. Starting with West Bengal, the governors themselves, namely the forest officials in charge of their vast landed estate, realized that old methods of control and exclusion were merely fuelling social conflict. An overworked and underfunded bureaucracy then started, slowly, to involve communities in forest working. What started as a strategic imperative became, at least for some forest officers, a sincere change of heart.

Once the critics from without were being echoed by the dissidents from within, the process of reform accelerated. This is indeed the signal lesson of Indian forest history-that meaningful policy change comes about only when the sustained pressure by social movements and their intellectual sympathizers resonates with the feelings of powerful officials within the state bureaucracy.

One or the other, by itself, will not do. When Brandis was active, he was handicapped both by his lone dissident voice within the Forest Department, and by the fact that there had not yet emerged an effective critique from outside. When Elwin, Mira Behn, and others propagated the feelings and aspirations of the peasants and tribals they worked with, the forest bureaucracy was, collectively and to the last man, deaf to their arguments and entreaties.

Forest policy remained unbending and unchanging, with the exception only of the Kumaun hills. There, as we have seen, the popular protests and outside critics were partially successful not because of a honest rethinking by the state but by its concern that this sensitive border region must not be tempted into outright rebellion. Elsewhere, where this political imperative did not come into play, the colonial regime refused to heed the widespread criticisms of its system of forest exploitation.

In more recent times, however, the radical critics have been aided by the autocritique of influential sections of the forest establishment. This confluence of external pressure and internal rethinking explains why, and how, the contemporary proponents of community forestry have, unlike their predecessors, been able to see their ideas and polemic become translated into official policy and (though less assuredly) into official practice. Nonetheless, there are indeed striking parallels between the ideas underlying the application of joint forest management today and the ideas of the early, prescient, and brave but for the most part unheard critics of state forestry discussed in this essay. With respect to the role of forest dependent communities, for example, there is a shared faith in indigenous knowledge, in the management capacity and robustness of local institutions, and above all, a sharp focus on local access to the usufruct of the forests.

Again, with respect to the role of the state, there is a common recognition of the essentially advisory role of the forest department, of its need to collaborate with rather than strictly regulate customary use, and of the justice of sharing revenues from forest working with the villagers.

Then, and now, critics have called strongly for an attitudinal change among state officials, a retraining and retooling in keeping with the democratic spirit of the age. Finally, both past and present proponents of decentralization seem to converge in their larger vision for forest policy in India, a vision which in my understanding consists of three central principles: (1) that benefit sharing (between state and community) and local control are to be the key incentives to ensure sustainable management and minimize conflict; (2) that community-controlled forests would work as a complement to a network of more strictly protected areas, further from habitations, that continue under more direct state control; (3) and finally, that the restricting of state control to these latter areas is vital on grounds of equity (i.e., the respect for local rights and demands), efficiency (i.e., as the most feasible course, with the state not biting off more than it can chew), and stability (i.e., as the most likely way to lessen conflict).

There is little question that the ongoing attempts at reversing or mitigating state monopoly over forest ownership and management do constitute a significant departure from past trends. In a deeper sense, however, contemporary attempts at fostering participatory systems of forest management hark back to a much older tradition. In the late twentieth century, as in the late nineteenth century, there has arisen a movement for the democratization of forest management, for a system founded not on mutual antagonism but on genuine partnership between state and citizen. The first inspector general's vision for Indian forestry was abruptly cast aside in the 1860s and 1870s, but it may yet come to prevail. That would be a vindication of the life and work of Dietrich Brandis, but also of Jotirau Phule, Verrier Elwin, Mira Behn, Govind Ballabh Pant, and the Poona Sarvajanik Sabha. Ramachandra Guha is an author and columnist based in Bangalore. His books include The Unquiet Woods (first published 1989; second revised edition, University of California Press, 2000), and Environmentalism: A Global History (Addison Wesley Longman, 2000). He is now working on a history of independent India.

6

Using Benefit-Cost Analysis to Evaluate Forest Management Strategies

Forest management expenditures, which are made in the expectation of creating benefits at some time in the future, are best viewed as investments. Although some foresters are reluctant to take this perspective, there are compelling reasons for doing so.

This opens up opportunities for using standard investment analysis techniques to rank projects competing for scarce funding. The use of these techniques lets forest managers evaluate alternative strategies and select the one which will contribute the most to the achievement of management goals. This technical note is intended to provide forest resource managers with an introduction to benefit-cost analysis and its potential use in forest management. We present a basic introduction to some important concepts of benefit-cost analysis, then discusses the nature and valuation of benefits and costs commonly found in forestry, and finally concludes with an illustrative example.

NORTHEAST SCIENCE & TECHNOLOGY

The basic purpose of benefit-cost analysis (BCA) is to assess the economic attractiveness of an action or project. A single project may be assessed to see whether it is profitable or meets a minimum profitability standard. BCA is also commonly used to rank alternate projects and select the most attractive option. Project assessment is based on a comparison of the costs of the project against the expected benefits. There are several ways in which the costs and benefits can be compared, but all of these measures share certain key characteristics.

In forestry projects, the benefits and costs are often incurred over a long period of time and they have to be discounted to allow for a consistent comparison. Discounting converts the value of future benefits and costs to today's dollar values. For someone with a discount rate of r percent per annum, today's value of $100 payable to that person in one year can be calculated as:

Present Value = $100/(1+r/100)

When r is ten percent, the present value equals $90.91. Looked at it in another way, $90.91 today can be seen as the present value of $100 payable one year in the future because $90.91 invested at ten percent will yield $100 in one year. Discounting is necessary because it is not accurate to treat equally dollars that are received or paid out at different times.

There are three measures commonly used to compare the costs and benefits. The approach that most economists favour is simply taking the difference between the total benefits and total costs. The result is known as the present net worth (PNW) or present value of a project or action. It is the net gain of the action. The formula may be written as follows:

(1) PNW = Benefits-Costs

When one is comparing alternate projects, the project with the highest PNW is the most desirable because it provides the maximum net gain. A project with a PNW just breaks even and a negative PNW implies a net loss from the project or action.

An alternative ranking approach is to calculate the value of the benefits derived per cost dollar. This is known as the benefit/cost ratio (BCR), and may be written as:

(2) BCR = Benefits/Costs

The project with the highest BCR yields "the biggest bang for the buck" and is the most desirable. A break-even project has a BCR of one and a money-losing project has a BCR of less than one.

The third common ranking criterion is the internal rate of return (IRR), which is defined as the rate of discount that produces a PNW of zero. Again the highest IRR is associated with the best project.

Of these three approaches, the IRR will not be explored further, since it can be difficult to calculate when the investment period is very long, as is often the case in forestry. The PNW is often the best criterion for ranking projects, but the BCR is a measure that can be useful in identifying forest management alternatives that lead to higher PNWs.

Benefit-cost analysis can be used to evaluate forest management at various scales. At the smallest scale, BCA can be used to compare regeneration alternatives for a particular site. At a larger scale, alternative regeneration strategies for a management unit might be compared. For example, there is much debate over the degree to which intensive forest management should be practised and BCA can be used to compare the benefits of pursuing strategies based on different management intensities. BCA can also be used to allocate funding among management units, or competing projects.

Identifying and measuring the benefits and costs is generally easier on smaller projects. The most common benefit of forest management is an increased volume of timber, either per hectare or from a management unit. Usually, the application of management changes the species that are found, especially on reforested areas, and it may lead to the production of higher quality stems. Management may also lead to increased population levels of wildlife species, increased or improved recreational opportunities, better regulation of water flow, and improve the aesthetics of an area.

Typical costs of forest management are expenses for silviculture, pest and fire protection, access and harvesting roads, and overhead. If a project is difficult to finance or can be expected to lead to cash flow problems, the cost of financing should also be included. Other costs or disbenefits may be incurred if wildlife populations or recreational opportunities are reduced, if the aesthetic quality of an area is reduced, or if water regulation is disrupted. Measuring the size of these disbenefits is often difficult but is necessary if they are to be valued. Some estimation will usually be required. Clearly, poorly planned and executed forest management activities generate costs that can be avoided.

The analyst must also select a discount rate. There are two chief options: the "real" discount rate (which has inflation netted out) and the "nominal" discount rate, which includes inflation. The advantage of excluding inflation is that there is no need to estimate future inflation levels and adjust future costs and benefits accordingly.

That leaves the analyst with the task of selecting an appropriate real interest rate. If the prime interest rate is denoted as r, the inflation rate as p, and the real interest rate as i, the relationship between these variables is given by:

(3) $(1 + r) = (1 + i) * (1 + p)$

Setting $r = 0.07$ (i.e seven percent) and $p = 0.02$ (i.e. two percent) results in a real rate of interest equal to 0.049 (4.9 percent). This is in the middle of the four to six percent range that is usually used for real interest rates. The valuation of these costs and benefits is more problematic. Increased timber production is perhaps the easiest benefit to value. The important point is that the benefit of a management action is the increase in timber volume, quality, or both. On Ontario management units, a low level of silviculture will not eliminate timber production altogether. Even when natural regeneration is used, there will be some timber produced. The value associated with implementing artificial regeneration is seen in the increased and more rapid production of timber, and often in the regrowth of more valuable species.

Timber can be valued at an amount equal to Crown dues, although Crown dues are not closely linked to factors that affect timber value. A more appropriate approach is to use the net value created by processing the timber. This is calculated by taking the market value of the products obtained from a cubic metre of wood and subtracting the costs of harvest, transporting the wood to the mill, processing in the mill, and transporting the product to market.

A study funded by MNR's Forest Values Group used this approach to value timber in Ontario (See SSK Inc., 1993)

In districts where tendered sales are common and representative of the local timber, the bid amounts provide a basis for estimating the value of standing timber.

The analyst must be careful to ensure that the BCA has the proper scope. Stand-level analysis is appropriate for landowners who are interested in single stand management and even for larger ownerships such as conservation authorities which do not require an annual timber harvest. Furthermore, stand level results can be usefully incorporated into forest level harvest scheduling programs when a relatively small proportion of the management unit area is affected by the treatment alternatives. However, if a manager is comparing treatments that will be applied to a significant number of hectares in a forest, then a forest level analysis is needed to properly account for between stand interactions. This type of analysis can be very complex and envolved and usually requires the use of computer modelling.

The intensification of a regeneration regime should yield a merchantable stand sooner than would otherwise be the case and presumably will contain a higher yield of a more valuable species. The benefits then will be the increased value of the future stand due to the higher stocking to the desired species and the earlier avialability of the areas for harvest.

Typically, the point at which a stand-level analysis begins is after the removal of the current stand. The rationale for this is that the money spent to regenerate the site (i.e. the investment) is related to objectives concerning the future stand, and it is on those merits that the regeneration expense should be evaluated. When spending more money on the harvest operation will reduce the costs of subsequent regeneration, the extra expense should be included as a regeneration cost, since this is the reason for it.

In a single stand analysis, one should ensure that there are no inconsistencies in the analysis caused by evaluating different alternatives over different time periods, as can happen when different actions will affect the stand rotation age. This can be avoided by examining a long series of repetitions of each alternative, effectively giving rise to an analysis period of infinite length. Alternatively, one could choose an analysis period based upon multiples of the harvest ages (e.g. compare two fifty-year harvest cycles under intensive management versus a one hundred year harvest cycle under extensive management). However, it is not always simple to find the required multiples, especially when a large number of options are being compared. This type of analysis is accomplished by considering the present net worth of a series of harvest values predicted infinetely into the future.

The value of wildlife is more difficult to assess. Forest management will likely shift the distribution of wildlife species in a stand or forest. Ideally, one would like to have estimates of how wildlife populations will be affected by alternative management actions. This is difficult, due to the number of species involved and the variations that populations exhibit over time. It is not yet possible to track all species-only those of importance or value to humans are tracked (although the criteria for importance are changing rapidly).

At the present, valuation of a major species poses significant obstacles. Suppose the moose population is locally important. Moose have value because people like to view them and because people like to hunt them. In theory, it is possible to survey people who live in and/or visit the area of interest, ask what value they place on having various population levels of moose, and adjust the results to account for the entire local (human) population and tourists. Statistical refinements based on income and level of camping or hunting experience are often made.

Studies in the U.S. have shown that a typical person will be willing to pay some amount for higher populations of many species because there are better wildlife viewing opportunities and because there is some satisfaction in knowing that there are more (rather than fewer) animals of that species alive in the area. A hunter will also be willing to pay some amount to have more game species in an area because the chances of a successful hunt are increased. A larger game population will also be likely to lead to an influx of hunters from other areas, which is a plus in a regional analysis but of little effect in a provincial or national analysis.

The author is not aware of any published data based on willingness-to-pay surveys that have estimated the value of wildlife species in Ontario. Where game species are involved, one alternative is to use the value of expenditures made by hunters as a proxy for the hunting value per animal. This has the obvious drawback of ignoring the value of benefits unrelated to hunting, but since expenditure data are available, they represent the best alternative for now. Of course, the benefit-cost analysis must still relate population level to degree of hunting effort and this will require some estimation.

The valuation of recreation opportunities poses similar problems. However, the key steps are the same as in the case of wildlife and timber. First, a quantitative relationship must be established between alternative management actions and recreation opportunities. Secondly, the value of a recreation opportunity, usually measured in terms of a recreation-day, must be estimated. One estimation technique is the willingness-to-pay survey. A more common technique, which is being employed by the MNR Forest Values Group, is known as the travel cost method. The central idea is to identify where the majority of recreation users come from, divide this region into zones of known distance to the recreation site, estimate the cost of travelling to the site from each zone, and then use the incidence of recreational use per zone and the zonal population to construct a demand curve for recreation at the site. Using site entrance fees is not appropriate for valuation purposes because prices are often set at low levels to encourage use. Finally, the valuation of such features as aesthetics is in its infancy, especially in Canada. Willingness-to-pay surveys of various types are again used in this case. Unfortunately, there are no known credible estimates of the value of aesthetic quality in Ontario.

There are two general classes of costs: out-of-pocket expenses and indirect losses or disbenefits. The out-of-pocket costs of undertaking various activities are usually unambiguous, but can be difficult to track down. Some judgement has to be exercised in the case of expenditures for fire protection and road construction and access. In these cases, a clear understanding of the reasons for making these expenditures is an important starting point. Fire protection is intended primarily to prevent the loss of human life, and secondly to prevent the destruction of buildings, machinery, or other capital goods. Therefore, only a fraction of fire protection costs can be charged against forest management.

At a forest level the choice of the exact fraction of the fire program's costs to allocate to forest management would depend upon the fire district, proximity to habitation and recreation values, wildlife concerns and of course the value of the forest and its contribution in providing wood or forest land supplies. Any attempt at a local division of these cost allocations would be artifical due to way these costs are accounted for at the aggregate program level, the interdependency of the social values associated with the use categories and the fact that in most areas fire prevention or control expensess are incurred to protect several use categories at one time.

Similarly, roads are built to facilitate timber extraction, subsequent assessment and regeneration, and provide access to the public for other purposes. Again, a subjective estimate of the proportion of the road value attributable to timber extraction is usually made. The costs of roads must also be amortised over some period of time, such as the 20 to 25 years often used for large capital expenditures such as pulp and paper mills.

As was mentioned above, costs of damage or reductions in non-timber values should also be included. These can be estimated using the principles that are appropriate for estimating non-timber benefits, described above.

The valuation of costs and benefits will depend on who is doing the analysis. A planner working for a Forest Management Agreement (FMA) holder will have a different perspective than a government planner on the benefits and costs associated with alternative management strategies. For example, since some regeneration costs are partially refunded by the government on FMAs, the company and the government will naturally place a different cost on regeneration. This is as it should be. A simple example will be presented which illustrates the use of benefit-cost analysis to compare the application of three forest-level strategies on a management unit. Suppose that these strategies are characterised as:

- Reliance on Natural Regeneration
- Extensive Management
- Intensive Management

The particulars of the strategies are unimportant; the key point is that each increase in management intensity results in a greater volume of wood being produced from the unit. Suppose that the wood supply levels are expressed as annual volumes. The table below shows, for each option, the annual wood supply, the value of the annual harvest, and the annual cost of management. Note that wood has been valued at $10/m3, and non-timber impacts have been excluded.

It will be assumed that some harvesting will be done, and so one of the three options must be selected. The key issue is the comparison of the net gains from management intensification. The table below shows the PNW and BCR associated with each option.

A comparison of PNW's shows that the greatest net benefit is associated with intensive management. Therefore, if sufficient funding is available, intensive management would be preferred. However, the BCR declines as management is intensified from extensive to intensive levels. This indicates that the gain per dollar spent decreases as the management level is increased.

If there are other potential investments that yield a BCR ranging from 1.46 to 1.60, then the greatest net benefit will be obtained by adopting extensive management and investing the remaining available funds in these other projects.

In this way, the BCR acts to guide investment so that the greatest overall PNW can be obtained. In this example, the profitability of management was at least partially justified by the increase of current harvest made in the expectation of future yield increases due to management. This is the allowable cut effect (ACE). Economists observe that the ACE has often been used to justify poortreatment investments (see Davis and Johnson, 1987, Pearse, 1990). It is considered that there are really two separate decisions involved-the selection of the harvest level and the selection of the appropriate silvicultural treatment. Mainstream economists argue that these decisions should be analyzed separately on their own merits, whereas forest economists point to sustainable yield requirements as the basis for linking the two decisions (Hyde, 1980).

In Ontario, the argument is clouded further by the public ownership of forests and the flow of non-timber benefits that come from them. This implies that the government should include in the analysis the social benefits that can be expected to arise from an increased harvest level.

What conclusions can be drawn from this debate? The first is that one should be reluctant to include the ACE when analyzing projects on private land where the sole purpose of investing is to grow timber but ACE is a valid factor to include in analyses of public forest management.

The second point is to heed the economists' warnings and critically examine the merits of the silvicultural projects being considered. Foresters should be careful about planting on medium and poor site classes, and on sites distant from the mill. Due to the additional costs of hauling it to the mill, the value of timber grown further from the mill is lower than the value of equivalent timber close to the mill.

One of the great difficulties in forest planning is dealing with the long time periods required to grow timber. The fastest growing boreal stands may be harvestable in 40 to 50 years and it is difficult to forecast real timber prices that far in the future with any accuracy. The uncertainty inherent in predictions for 100 years or more is enormous. Therefore, in view of the significant degree of uncertainty associated with assessing the profitability of a regeneration project, one may question whether such an exercise is worthwhile.

Nevertheless, most people believe that planning is useful and planners can only use their best estimates. It is also the case that accurate forecasts of future prices and costs are less important when one is using BCA to select the most attractive project or to allocate funds among projects.

In these situations, any errors in forecasting will apply equally to similar projects and reservations about long-term forecasting are less relevant.

Forestry debates present society with several choices that need evaluation. The task is to estimate the value in each of the choices presented. The purpose of this paper is to establish a framework that will allow each environmental policy option to be systematically and comprehensively assessed for its impact on the total economic welfare of the community.

Such a framework consists of ensuring that the non financial values, financial values and alternative financial values derived from the forest are accurately evaluated, and hence given their full and due consideration.

There are presently significant barriers to the achievement of this objective. The three barriers identified in this report are:

1. The existing inability to determine the correct financial values for forestry operations.
2. An inability to determine the correct financial worth of alternative activities such as plantations and water production.
3. The lack of a methodology which determines the relative worth of non financial values (NFV's) of the forest resource.

Removal of these impediments, which is an exercise in microeconomic reform, will bring improvements in the financial efficiency of the Australian economy, whilst improving the overall quality of life for Australians in general. The first two impediments can be removed by the utilisation of some fairly common methods. With specific reference to the third impediment, however, there are many approaches by which appraisal of non financial values can occur.

Some of these are specific methodologies such as contingent valuation and public ranking of options. Other methods centre around public involvement in the assessment of options. Yet other methods are based on using 'rules of thumb' to account for non financial values.

Among the reports surveyed, some techniques are mentioned quite frequently. These included public participation as a key necessity for getting non financial values properly assessed. Surveying of public opinion, systematic reviews of natural features and the use of experts also seemed to be relatively common.

The task therefore becomes one of finding an appropriate route through the range of techniques offered.

Section 3 of this report briefly outlines the relationship between the concepts of externalities, economic value and community well being. The absolute requirement of good economic analysis is the need to accurately appraise all values, including externalities (non financial values) in any assessment framework. Economics provides the theoretical standard by which the existing assessment frameworks can be judged.

Section 4 reviews several major forest reports in order to identify four significant components of forest project appraisal. These include the values of the forest, the valuation methodologies proposed for assessing values, the pricing/ financial valuation of forestry and alternative activities, and the assessment framework for appraising policy options in the search for the socially best outcome.

In particular, Section 4 examines the role of pricing in the appraisal of forest policy options. The existing assessment processes and the various forest reports (apparently on the grounds of practicality) adopt a 'compartmentalised' approach to pricing. In other words, protection of non financial values is carried out prior to, and separate from, the pricing of forest timber. If the appraisal process is incomplete, then the pricing of timber is inadequate by definition. Pricing policy is significant for estimating value to the extent that it partly determines the financial value of forest timber.

Section 5 then describes weaknesses in the existing appraisal processes. These flaws range from a broad neglect of non financial values in project assessment to more specific issues. These include the failure to comprehensively identify and value non financial values, often even in the form of physical impacts; failure to identify or consider alternative financial values and alternative project options; and the failure to provide accurate statements about the financial viability of projects.

This report makes clear that there is indeed evidence of serious weaknesses in the determination of the financial values of native forest timber. Our analysis indicates the very strong likelihood that native forest timber harvesting, as a commercial activity, is being subsidised (eg. low cost public sector loans, no commercial charges for land), is inefficient (eg.. there are better returns to be had from harvesting water in at least one case) and poorly managed (eg. the agencies are unable to provide appropriate accounting information on their commercial activities).

Consequently, a significant microeconomic reform process needs to be undertaken, if the correct financial values for native timber sales are to be determined.

Inappropriate appraisal of financial values may also be suppressing economic growth in the private forestry and agroforestry sectors (and other alternative/substitute financial values). The lack of a 'level playing field', with respect to the cost of business inputs (eg. land and capital) between the public and private forest managers, suppresses investment in private forestry at a lower level than that which could normally be expected.

If financial values and alternative financial values of the forest were to be appropriately appraised, this may identify alternative revenue streams and investment opportunities for hard pressed forestry commissions.

Summarised, inaccurate financial values and a limited knowledge of alternative financial values combine to undermine the basis of comparison between the financial and non financial values of the forest. This problem is compounded by the lack of a methodology for determining the relative worth of non financial values.

In conclusion, it must be reiterated that any assessment of the social benefits and costs in forest policy must include three pieces of vital data: an accurate appraisal of financial values, alternative financial values and non financial values. Without reference to such data, environmental decisions become less an attribute of good management, than of reliance on sheer blind luck.

Resolving these problems will contribute to better environmental management, greater financial performance and more jobs. This makes reform in this area a key part of the process of microeconomic change, resulting in better environmental and economic policy.

7

Future Forest Conservation Policies

It is stating the obvious to note that the native forest estate is increasingly the centre of considerable public controversy. The debate over forests is a microcosm of the broader debate surrounding environmental issues because it continually confronts society with complex choices.

Each choice offers a variety of paths to the future, and in turn, each path provides net 'value' to society. Net 'Value' is comprised of a mix of quantities and types of 'values' which individually and collectively meet the needs of society. Differing 'paths to the future' confront society with the need to estimate the net value likely to be derived from each route. In economic analysis, and (it is assumed) society at large, the path which offers the most value to present and future citizens is the preferred choice.

Many complex social choices are decided in the realm of politics with some choices arbitrated on a more technical, arms length basis by designated officials. Choices in environmental policy historically combine technical, 'objective', elements and qualitative, 'subjective', elements. As a result, the choices are subject to a range of pressures that may not beset other types of decisions. In this context, it follows that estimating the value, or total 'economic welfare', of different paths within environmental policy may be more difficult than the simpler financial trade offs common to economic decision making.

'Optimising the benefits to the community' has been identified as a key objective in forest policy. In order to achieve this goal, it is necessary to identify the contribution that different policy options can make to the welfare of the community. In other words, we must determine which path to the future offers the greatest 'net value', and which combination of values acts to create that path.

This discussion examines the relationship between financial values derived from timber harvesting, the financial values derived from alternative uses of the forest (eg. water, tourism etc), the financial values derived from plantation timber and the non financial values of the forest (eg. environmental amenity) through the reviewing of a series of reports on native forests.

THE CONTRIBUTION OF ECONOMIC THEORY

The Ecologically Sustainable Development Working Group on Forest Use specified three main requirements for sustainable forest use:

- maintaining the ecological processes with in forests (the formation of soil, energy flows, and the carbon, nutrient and water cycles);
- maintaining the biological diversity of forests;
- and optimising the benefits to the community from all uses of forests within ecological constraints. The National Forest Policy Statement adopts these principles as the basis for ecologically sustainable development. (National Forest Policy Statement (NFPS), 1992:47) (Emphasis added.)

Economic theory sets a standard by which the outcomes are measured, and informs us as to what are 'good' and 'bad' ways of making choices. Economic theory also provides an important underlying philosophy which specifies that changes should be to the net benefit of society at large.

The Role of Economic Theory

Economic theory is often seen as a method, or science, of analysing impacts on the financial e c o n o m y. When properly and rigorously deployed, however, it is elevated to the 'art ' of choice. Thorough economic analysis concerns itself with how to make choices that provide the greatest net benefit to the community, where 'benefit' implies consideration of all social values. It is for this reason that economic theory, incorporating such concepts as externalities and public goods, has come to play a significant role in decision making processes. The following sections briefly outline the relationship between the concepts of externalities, economic value and economic decision making.

Externalities and the Welfare of Society

Economics is centred on the allocation of scarce resources to competing ends, in order to achieve the most value for human society. In this instance, economic analysis seeks to identify the flows of value that can be derived from 'uses' and 'non uses' of the forest.

This process of analysis seeks the combination of uses and non uses that most meets the value preferences of society, both now and in the future. The process is intended to be value neutral in the sense that it is based on the given or expected values of society-whatever those values may be.

Traditionally, economic theory uses the 'market' as its benchmark of efficient resource allocation; that is, an allocation of resources that creates the greatest benefit for society. This 'pure' model, however, admits that its results hold true only under very tight and specific conditions. If any of these conditions are breached, then there can be no presumption that the free market outcome is the best one.

Externalities arise when certain conditions of a free market do not hold. They arise when the free market fails to include, or take account of goods and services provided incidentally by the completion of another transaction. An externality can be compared to an auction where the main transaction is completed between two parties, but where additional goods and services, not provided for in the original transaction, are subsequently delivered.

These can be delivered either to the main parties or to third parties who were not involved in the original deal, and who, indeed, may even be unaware of the original transaction. For example, if a tree is cut down in the forest for timber, the loss of that tree may adversely affect the view of a home owner, who subsequently experiences a loss in his/her property value.

Externalities and public goods are both expressions of market failure. 'Public goods' refers to a value from whose benefit others cannot be excluded. Thus, deriving pleasure from knowing of the existence of a forest is an externality, because others cannot be excluded from deriving the same pleasure. At the same time, the phenomenon is also a public good because it is non rival in consumption; that is, many can consume the item without reducing the consumption of others.

In order to avoid confusion, all such values are called 'non financial values' since they cannot, by definition, be traded in a market place, and hence be expressed in terms of financial value. Where environmental issues are concerned, the market cannot be relied upon to produce an optimal allocation of resources-ie. the greatest overall benefit to the community-because there is no venue, no 'global auction room', in which the community (in particular future generations) can express its preferences over the items it values.

The 'items of value' in this discussion are the 'values' or 'benefits' that flow from the 'use' of 'objects'. Normally, economic analysis talks of objects such as goods and services when the real interest is in the values that flow from these objects. It makes little sense to talk, for instance, of a 'forest' per se. Are we referring to the wood alone, the biological species that reside there, the pleasure that bushwalkers derive, or its function in absorbing greenhouse gases? The term clearly has no operational meaning unless we refer explicitly to its various use values and benefits. In order to avoid any ambiguity, the remainder of this paper will generally use the term 'values' rather than 'goods' or 'products' (as 'values' is a more inclusive term).

An understanding of externalities is central to quality economic decision making, particularly in the context of the environment-simply because in the real world, there are many items of value (fresh air, knowing that an old growth forest exists, etc) that cannot be bought and sold in the market. The exclusion of these items from the 'economic equation' means that the overall equation is miscalculated. Miscalculations of the total economic equation generally lead to a misallocation of economic resources, and to an overall loss to society.

The practical significance of identifying externalities and public commodities relates to their impact on markets. It is impossible for a market alone to systematically produce the socially optimal outcome when externalities and public commodities are present, as they are in the case of the natural environment 1.

Pricing Theory and the Financial Values of the Forest

In economic theory, the price of an item can represent one of three circumstances for the owner.

- The price could reflect the value of the item to the owner.
- The price could value the item at a level greater than the owner's subjective valuation. In this case the owner is likely to sell the item.
- Alternatively, the price of the item may be less than the value to the owner in which case the owner will refuse to sell.

The price of wood must also reflect the different type of values generated by the wood and the portion of forest which is altered by its extraction.

Hence, if the 'public citizen' 2 offers a logging coupe for sale, by definition the value anticipated from the sale must at least exceed the value expected from any alternative use. And additionally, the price needs to equal:

- the costs of the production of wood;
- plus a return on the capital invested, with sufficient reward to cover the risk; and
- total compensation for all non financial values (NFV's) lost in the process of an extraction of the timber.

This pricing method ensures that timber with values exceeding the market price will not be sold. However, the significance of

NFV's is that they make up a considerable proportion of the value of timber. When NFV's are destroyed, the loss of value is not felt by one person alone, but is likely to be felt by every Australian citizen, every international citizen and everyone yet to be born. When logging occurs, each of these people experiences a change in the value they derive from the forest, and hence each should be compensated. The price of timber needs to be able to compensate all losers if policy is to be socially optimal.

The pricing of timber is crucial to determining the pattern of logging in the forest. In order to implement a solution, the present approach of forestry commissions uses regulation to separate logging areas from non logging areas. This approach could be assuming one of two things. Firstly, it could be assuming that royalties are sufficient compensation for the loss of NFV's 3. In this case, it remains to be demonstrated by those who set royalty levels that they have evaluated the worth of the lost NFV's, and thus set royalty levels accordingly. In the absence of information about how NFV's are determined, it could be assumed that non financial values in the logged areas are considered insignificant to the extent that the Australian community will not require compensation for the losses in NFV's which result. In theory, logging will then occur, in the areas set aside, if the cost of production and capital makes it a financially rational exercise.

Pricing policy is, thus, conducted on a 'compartmentalised basis'. In other words the forest is surveyed for its values, and those areas where NFV's are deemed 'small enough' to avoid the need for compensation, are made available for logging. The remainder is put aside for conservation. Thus, if we use a 'compartmentalised' approach, timber is valued in a two stage process. The forest is allocated to different uses depending on the scale of NFV's in different areas. Forest that is to be subject to logging is then priced on a commercial basis. Pricing takes place on the assumption that the coupes have already been surveyed, and the NFV's have been found to be minor; ie very close to zero.

The worth of the royalties, of course, must also cover the costs of wood production, the return on capital required, and the financial opportunity cost of foregone uses. The Victorian Auditor General's report (1993) indicated the likelihood that the forestry operations in Victoria were making low profits (if any at all). Given that royalties must cover operating, capital and opportunity costs, this leaves very little room to compensate for lost NFV's. In such a circumstance, it is feasible that NFV's are being valued at a worth very close to zero. No evidence was identified to justify such a valuation.

THE THEORY OF TOTAL ECONOMIC VALUE

In any public decision over the net benefits of forests, externalities and public commodities, along with project and alternative financial values, need to be taken into account, as part of the value provided by a 'resource'. In analysing different resource options, a total listing and weighting of values, both private and 'externalised', must be provided for each option.

> When an economist is comparing alternative social arrangements, the proper procedure is to compare the total social product yielded by these different arrangements... In devising and choosing between social arrangements we should have regard to the total effect. This, above all, is the change in approach which I am advocating.

Coase refers to 'total social product' which we have renamed 'total economic value' in the tradition established by Pearce (1989). In order to compare the total economic value of differing policy options, it is necessary to aggregate for each option all the values (financial, aesthetic, spiritual, etc.) that would be produced for the community if a particular policy option were to be undertaken.

Traditional project evaluation normally compares one option against another option, and chooses that which has the most value. In the presence of market imperfections such as externalities, the total economic value (the project's total costs and benefits to society) can be severely miscalculated in at least two ways. First, proper economic evaluation needs to consider the financial impact that one project may have on other parties, not simply whether it is profitable when considered in isolation. Suppose, for example, that the government is requested to give its approval for the establishment of a factory on a river bank. Viewed in isolation, the proposal is attractive: it is expected to provide jobs for 100 local unemployed young people and will contribute an extra $5 million to tax revenues. Viewed from a wider perspective, however, the project's net financial value would be seen to be negative if the poisonous effluent it emits cause the bankruptcy of downstream fish farms and tourist resorts which together generate 200 jobs and $10 million in tax revenues. These 'down stream' financial values are often overlooked by traditional project evaluation methodologies.

The second weakness of the traditional evaluation approach is that it gives little consideration to non financial values. To continue with the illustrative example above, the factory's pollution may poison wildlife and household pets, ruin a picnic spot favoured by thousands of local residents, etc.

These types of costs are real and relevant, notwithstanding the fact that they are of a 'non financial' nature. Resolving the first problem is easier, since it may be possible to obtain the relevant financial data if appropriate economic modelling is conducted. The second problem is more difficult because non financial values are not traded in markets. In other words, there is no 'objective dollar price' which can provide a yardstick with which to compare non financial and financial values. Environmental economists have been working on providing techniques to evaluate non financial values.

Economic Decision Making

As a society, and as individuals, we are constantly evaluating and rejecting options. A social cost benefit analysis is, in effect, the act of making choices. Social cost benefit analysis becomes a conceptual framework that expresses the less formal processes we use to make choices. While the National Forest Policy Statement leaves open the practical question of assessment, it identifies the need for evaluation of policy options. In Diagram 1 we have listed all the values, in the tradition of Pearce (1989) and Young (1992) that appear to be provided by the native forest. Table 1 describes each type of value, and where relevant, their relationship to each other. Total economic value (TEV) analysis would require that each policy option would receive a 'score' in each cell of an evaluation matrix (see Table 5 for an example). Scores are aggregated by a means yet to be determined which then enables decisions over differing options to be made.

Economic decision making seeks to choose between the differing options on the basis of determining that which provides the greatest net social benefit. The means of deciding which option has the greatest net social benefit where 'scores' cannot be aggregated into a single numeraire is, of course, highly contentious.

There are three channels through which resources can be allocated. These are the markets, institutional processes and executive decision making. Markets, however, as decision making institutions, are largely unavailable in the circumstances covered in this report because we are dealing specifically with instances of market failure. However, modified markets (ie. markets subject to government intervention) are included as a means of policy implementation. The choice between government modified markets and 'pure' market measures should be justified on the grounds of net social benefits.

The irony is that if market measures are to be chosen as the decision making process, then it is a decision that will have to be made by government anyway, and hence does not relieve the public decision making process of the need to evaluate the optimality of value flows from the forests under differing policy options. In other words, choosing to 'do nothing' is as much a decision as choosing to 'do something'. The lesson is that if optimal social benefits are sought, then market failure automatically triggers government decision making processes-even if that results in a decision to 'do nothing'.

The creation of a formalised process for considering environmental values and other externalities requires a clear understanding of the integrated structure of externalities, economic value and economic decision making. All values-financial, ethical, ecological, etc-must be accounted for, if different policy options are to be compared on a equal basis. The different values that comprise each policy option, need to be integrated in some manner which enables socially optimal-choices to be made.

APPRAISING THE VALUES OF THE FOREST: THE THEORY AS PRACTISED

This section reviews thirteen recent reports on forests for their views on the appraisal of forest values. Each report was reviewed in order to gain an understanding of the value assessment problems in the forest sector. The reports were examined on the basis of four criteria:

- Identification of externalities.
- Identification of valuation methodologies.
- Identification of pricing/financial valuation practices.
- Identification of likely assessment frame works.

Problems in the Identification of Externalities

A representative example, of the values identified, is listed below. These values were derived from the Resource Assessment Commission (RAC) study. This study, amongst those examined, was the most comprehensive and thorough in its treatment of the issues. The weaknesses in its study were repeated, and magnified, in other studies.

The Resource Assessment Commission Inquiry into Forest and Timber

Conservation; wilderness; recreation; Financial values such as wood, tourism and water supply, oils, seeds, tannins, medicinal plants, sandalwood, bark products, fishing, hunting, wild foods, infrastructure corridors; Information via education, research; intrinsic values like aesthetic (different from visual), heritage, spiritual, biocentric, wilderness, lifestyle; Utilitarian values; 'pristinity'; landscape; scenic quality, public sensitivity, industry value by implication; existence values, option values, vicarious use values, bequest values, ecological sustainability, ethical values.

The main weaknesses in assessing the manner in which values had been surveyed were that:

- Values were not identified or listed in a systematic or comprehensive manner. The above extract from the RAC report was amongst the most comprehensive, but even so, lacked a systematic listing of values.
- Different value categories often overlapped. For example, ecological sustainability was repeatedly confused with 'environmental amenity'.
- There was also repeated confusion over the values we impute to an object, and the values we derive from that object as a result of our preferences. For instance, a forest site may have great scientific value, but not be of great aesthetic value. Conversely, a forest site may have limited scientific value, but great aesthetic value. In the latter case, a purely scientific view point would not reveal the aesthetic importance of the site. Similarly, there are other values-for example existence value-held subjectively by the public with respect to native forests. A study which evaluated objective characteristics only would not pick up the subjectively felt values. We have identified this as the 'object subject' tension.

Confusion over the type and source of 'values' will undermine the efforts to assess all values. It is therefore necessary to have an agreed list of all values, comprehensively and systematically ordered, in order to achieve the goal of optimising benefits to the community.

The twentynine uses identified each provide seven different types of value which give their total worth. For example, use D 11 (craft industries) provides D 11.1 (direct use value), D 11.2 (option value), D 11.3 (quasi-option value), D 11.4 (existence value-sometimes known as industry value), D 11.5 (intrinsic value), D 11.6 (combination value) and D 11.7 (bequest value). Each of these seven components combines to give the total value provided by craft industries. In turn, the 'use' known as 'craft industries' can be combined with a package of other uses that will generate the total economic value of the forest:

While within a particular use, different components can be combined to provide a total value for that use-this does not imply that the-relationships between each component are necessarily positive. They may in fact be inverse for a range of values. Similarly, the relationships between different uses can be positive, negative or zero-even though total economic value is additive of the components.

In this report, each use has been ascribed a unique positioning (from D 1 to D 29) which is used in Table 1 and Diagram 1, and wherever these values are referenced in the text.

Table : Definitions of values

Value	Name.	Definition.
D1-D12.	Financial values	Equivalent to the net present value of any commercial operations that sell items produced by the natural processes of the forest.
D13.	Traditional Use values	Value derived from gaining a traditional, indigenous lifestyle from the natural processes of the forest.
D14-D15.	Unallocated.	
D16.	General information value.	The forest provides information that is of value to society at large. (This section is about information in general, not including biodiversity or mineral information.) An example of the information value of forests is the use of Tasmanian Huon pine in global climate change research (Quantum program, ABC TV, 1993)). This section largely incorporates the 'scientific' values which are the characteristics human society attributes to objects possessing great 'informational' value. The remaining values (or scientific values) derived from the characteristics of objects fit under the categories of cultural and heritage values. 'Scientific' values is used as a reference to all knowledge systems, including, but not limited to 'western science'.
D17.	Value of biodiversity information	Information about biodiversity is a subset of information in general. This information has been singled out has having particular significance to society at large and hence has been separated out.
D18.	Value of mineral resource information	Similarly information about mineral resources has been singled out as important and hence has been given a category of its own.
D19.	Recreational value	This is the value that recreational use of the native forest provides to the society.
D20.	Visual amenity value	The visual beauty of the native forest is a significant ambient value that is viewed directly or often on film or television. It is intended to include such terms as 'aesthetic' and 'landscape' for example.
D21.	Ambience value	Ambience is the general category that covers the ambience (experiential) that native forests provide to users. Visual amenity is part of this category, but

D22.	Cultural value	The native forest provides a source of cultural value to the present population either through indigenous culture or through links with traditions such as 'The Man from Snowy River'. In this sense the native forest helps to give present Australians a sense of who they are.
D23.	Heritage value	Heritage values represent the historical significance of the forest in Aboriginal and European histories and sense of perspective. The native forest is important because it may represent something about Australia's past.
D24.	Social value	The native forest provides a social ambience to the residents who live there, that may well be different from that which the same people would experience if their town was located in a different environment. Here the gain or loss of value is the net difference these people would feel between their present location and some alternative location.
D25.	Damage prevention value	The native forests are of direct use to society to the extent that they contribute to or mitigate the impact of natural phenomena on the human community and its activities.
D26.		Unallocated in this analysis.

D27-D29.	Ecological services value	The very physical existence of the native forest environment provides services to humanity labelled by this report as 'values'. The following three items are meant to comprehensively cover these services. These three items are intended to describe physical characteristics of the object called 'environment' that are valued of themselves. This comprehensive listing is only a first iteration. The breakdown of ecological functions offered in Diagram 1, could be replaced perhaps by the ecological attributes offered by the Australian Heritage Commission approach (AHC-CALM, 1992: Vol 1. pg9). Providing a testable measure of the ecological functioning of native forests is obviously the task of the relevant scientists and not the province of economics, though economic analysis suggests that the development of such a measure is absolutely necessary. The examples offered are only to stimulate the technically qualified to take the process further, or create a better method.
D27.	Ecological function value	Ecological functioning provides many services that underpin the existence of humanity and the condition of the planet generally. It is of value to our society that these services are provided and hence we have referred to this service as 'ecological function value. The NFPS has instituted ecological sustainability as a minimum standard beneath which ecological functioning should not fall. The ecological system of the native forest consists of many components that have been listed in Diagram 1.

D28.	Environmental quality value	The environment may be valued also, because it is of a certain level of physical quality. This quality may be closely related to the functional condition or it may be only partially related. The physical quality of an environment is of itself a direct service which society may choose to value.
D29	Disturbance quality	The environment may also be of greater value because it is undisturbed. An environment can be used in an ecologically sustainable manner, with good environmental quality, whilst being very disturbed by human activity. The level of disturbance of an environment may make a significant contribution to the physical services it provides. The above listed values D1 to D29 correspond to direct use or non use of the native forest. These are known as direct use values. The direct use, or non use, of the native f o rest produces value which economists have labelled direct use value. In addition to the direct use value, however, a further six types of value are produced. These are numbered D*.2 to D*.7 since direct use value is number D*.1. Thus with respect to the native forest, there are 29 identified uses (bearing in mind that numbers D14 & D15 have not yet been allocated). For each use, or non use, there are seven different types of value produced. Each of these must be taken into account in the TEV calculus. Thus use number D27 produces values D27.1. for direct use value, D27.2 for option value, D27.3 for quasi-option value, D27.4 for existence value, D27.5 for intrinsic value, D27.6 for combination value and D27.7 for bequest value.

D*.2.	Option value	Every value listed here also has an 'option' value. That is, option value is dependent on the existence of another value, if it is to exist itself. A value provides services directly, and hence it is also sometimes worthwhile to pay for the option of keeping that value available for future use, hence option value. The worth of an option is the amount willingly exchanged in order to keep an option open. (see Pearce and Turner, 1990 for further information)
D*.3.	Quasi-option value	This is what people would be willing to pay in order to have further information, on which to base decisions about the native forest. (see Pearce and Turner, 1990 for further information)
D*.4.- D*.6.	*are existence values*	Existence value is the benefit we receive by knowing that the existence of other values is secure. (See Pearce and Turner, 1990) There is an existence value for industry as well. This has been termed an 'industry value'. It is a value in excess (negative or positive) of that due to net p resent value. (Source Tasman Institute, Moran et al, 1991)
D*.4.	Existence value	Each value listed in this table can create a benefit in the minds of those who know of the existence of that value, and are glad of its existence. Ecological function value is of use directly, while knowing that it does exist is an additional benefit.

D*.5.	Rights of other species and the task of stewardship (including intrinsic value)	It is a philosophical view that other species, ecosystems and the earth itself, are of intrinsic value. That is they would be valuable even if humanity did not exist. The object (native forests) has rights of its own, and this is valued by the subject (individuals with this philosophical view) and hence is an addition to the worth derived from the object for other instrumental/utilitarian reasons.
D*.6.	Combination value See note below	Humans derive a range of values from the native forest. In combination some values are worth more than the sum of their parts. For instance, a beautiful view, combined with a high level of other ambient values may be more valuable than the two values when separated. The native forest comprises many components which each give rise to the values listed in this table. When those components are combined to form the 'native forest', the combination is of value in itself. Similarly a set of antique chairs is less valuable sold separately, than as a group. Secondly, combination effects can occur when different values are received simultaneously.
D*.7.	Bequest values	To be able to bequeath the native forest to the future is a value that many members of the present generation would experience. The benefit derived from this is known as 'bequest value'.

It is of note that 'Hoehn and Randall have shown that independent estimation of the component parts [of Total economic value], followed by their summation, is not formally identical to TEV estimated by a sequential procedure in which, say, existence value is estimated first, then 'site' experience value, then activity related values at the site.

This is because the consumer surplus for a bundle of goods is not the same as the sum of the consumer surpluses on individual goods.

Identifying Suggested Valuation Techniques

Each report suggested, implicitly or explicitly, a range of valuation techniques for providing an 'objective' assessment of the values identified. Overall, the lack of comprehensive identification of values, common to all thirteen reports, subsequently inhibited the matching of valuation techniques to appropriate categories of value.

This complicated the process of value assessment since it provided limited guidance about how various values should be measured.

On the positive side however, the Resource Assessment Commission (RAC, 1992) report did provide an assessment of each evaluation technique, which marks a very useful first step towards providing a listing of techniques relevant to the assessment of each value.

The Organisation for Economic Co-operation and Development references (OECD, 1986; 1992) also made a worthwhile contribution towards evaluating techniques.

In general, the reports provided only limited advice on the means by which different types of values should be evaluated. Moreover, with the exception of the RAC and the OECD reports, the techniques put forward were generally insufficient to measure the full range of values derived from the forest.

For example, the reports would discuss techniques for measuring the scientific values of the forest, with out discussing other measures that evaluate the 'worth' of the non-scientific values.

A second issue worthy of further discussion in each report was the role of uncertainty in complicating the assessment of values.

Whilst some discussion of uncertainty was entered into by four reports (RAC, 1992; ESD, 1992; AHC, 1992 and OECD, 1986), the issue was not raised sufficiently to provide guidance on how valuation techniques were to account for uncertainty. What is required is a more explicit discussion on the means by which 'uncertainty' should be handled in valuation processes.

The following table lists the valuation techniques that were identified. While there will be some debate about whether these are all 'techniques' or 'management strategies', it is clear that the implementation of these processes leads to the de facto valuation of forests.

For the purposes of this report, we have described them all as valuation techniques. Some of the techniques identified below were implicit in the various reports rather than explicitly described.

Table : Valuation techniques.

	Valuation Technique.	**Relevant Reports.**
1.	Regulation by arbitrary decision.	NFPS
2.	Regulation by identifying proxy variables for key values.	NFPS, RAC
3.	Regulation with 'rule of thumb' measures; eg. sustainability as a management criterion.	Most reports
4.	Ecologically sustainable management: so commonly used as a rule of thumb that it warrants a category of its own	NFPS, RAC, ESD etc.
5.	Conservation reserves: This technique implicitly values enclosed areas.	NFPS
6.	Community participation facilitates the expression of public preferences in decision making processes.	RAC, ESD, OECD, NZ Resource Management Act
7.	Participatory Rural Appraisal: a subset of community participation.	IIED, IDS
8.	Interest group roundtable.	RAC, ESD, OECD
9.	Community discussion panel: a group of informed local people ('imaginative extroverts': OECD).	RAC, OECD, ESD
10.	Public hearings.	OECD, RAC, ESD
11.	Visual management systems	RAC
12.	Balanced panels of experts	NFPS, RAC
13.	Referendums	OECD
14.	Focus groups & other techniques from marketing & advertising.	RAC
15.	Ranking of options	OECD, AHC

The following techniques did not appear in any particular report, but were the result of conversations with various experts in different disciplines.

This more multidisciplinary discussion revealed the following four measures as being worthy of further investigation.

17.	Use of video/audio techniques to capture relevant values of the forest. Based on use of pictures and film in advertising. Visual techniques enables the visualisation of values which may otherwise have to be described in a written form. In this sense, the decision making process is no longer dependent on third parties to value at least the visual aspects of the area	Suggested by experts from other disciplines.
18.	Comparative survey techniques: for ascertaining public responses to subjective values by comparing similar sites, to determine the direction and scale of public feeling. Valuation Technique. Relevant Reports.	Ditto
19.	Benchmark survey: asking the public to compare sites against a well known set of 'benchmark' sites.	Ditto
20.	Psychological research may provide some useful techniques which have not yet been tried before in this area. For example, Art Therapy is a psychological technique where patients are asked to draw pictures. This technique, among others, may have something to offer valuation methodologies.	Ditto

The OECD (1992:10) provided the following categories for the more familiar measurement techniques:

21.	Identification of surrogate markets (revealed preference approach) For example hedonic pricing seeks to elicit values by comparing property prices.	OECD, RAC
22.	Market 'creation' via questionnaire based techniques. eg. Contingent Valuation Method (CVM).	OECD, RAC
23.	Dose response data linked to valuation: eg. Preventive expenditures, Replacement costs, costs of restoration	OECD
24.	Public preference identification: 'Identification of public (social) preferences as revealed in social norms, rules, regulations and legislation' etc.	OECD, (OECD, 1992:11)
25.	Hypothetical/Indirect Approach to valuation: 'Individuals are asked to respond to hypothetical market situations, but their responses are only indirectly related to valuing the environmental good in question.' (OECD, 1992:11)	OECD

Further to this are the techniques used to assess financial values. These fall into two major categories in the reports reviewed in this study. The techniques are well known, and hence are only identified for the sake of completeness.

26.	Financial cost benefit evaluation of projects to ascertain the net present value of the investment.	Implicit in all reports, though often not well practised.
27.	Economic modelling/cost benefit analysis to ascertain the broader financial impacts of the project on the 'economy' and on other alternative financial values (projects).	RAC, OECD, World Bank

As we can see, the range of techniques identified was broad. They appear to have the capacity to address the measurement of all the values identified in Table 1/Diagram 1: However it is apparent from reading the various reports that some techniques are not fully accepted or understood. This lack of an agreed set of techniques, at the very least, complicates the process of valuation.

THE ASSESSMENT FRAMEWORK-A SUMMARY OF FINDINGS

Reviewing the various reports on the issue of forest policy has provided an insight into the process by which decisions are made. This section seeks to summarise the inadequacies that have been identified in the respective assessment frameworks and processes.

Australian & Overseas Experience of Assessment Frameworks & Processes

There are several assessment frameworks used in Australia for forest decision-making. Most of the reports have only considered assessment frameworks in general terms, without specific reference to real examples. In order to evaluate these assessment frameworks, the report provides a 'checklist'.

The 'checklist' provides a comprehensive series of questions which any worthwhile assessment framework should be able to answer. The 'checklist' approach facilitates identification of the broad strengths and weaknesses of existing or proposed assessment frameworks

A common theme in all reports (both Australian and international) was an underlying philosophical assumption that a 'technical', non-political choice process was possible, and to some extent desirable. This process was not challenged explicitly in any report, despite the obvious claim that the political arena is the desired place for settling value disputes (indeed it could be argued that this is the raison d'etre of politics). For the moment, however, this discussion accepts the premise that a 'technical' or 'black box' approach is feasible for value choices.

So as to gain an understanding of international experience, the report has sought to review processes suggested by the OECD and the World Bank. These internationally-recommended approaches provide a benchmark for achieving 'world best practice' in this field. The report has also examined the type of system deployed by New Zealand under the auspices of the Resource Management Act (NZ, 1991).

The international reports reflect a worldwide concern over the assessment of environmental values. While it is now recognised that environmental values must be incorporated, it is not known how this should be achieved. The World Bank view is not unusual:

Externalities are thus clearly troublesome, and there is no altogether satisfactory way to deal with them. This is no reason simply to ignore them, however; an attempt should always be made to identify them and, if they appear significant, to measure them. (World Bank, 1992:23)

This accurate, but unhelpful commentary is largely due to a neglect of environmental factors within economic analysis. This was noted by two internationally renowned economists in 1988, with respect to the field of development economics:

Environmental resources appear in this literature about as frequently as rain falls on the Sahara. (Dasgupta and Maler, 1990:102)

They also noted, as recently as 1988, that one of the world's best known texts on development economics (see note below)

... simply has no discussion of environmental resources and their possible bearing on the development processes.' (Das gupta and Maler, 1990:102)

This is symptomatic of neglect within the broad field of economic analysis. The OECD (1992) remarked that conventional cost benefit analysis would not include a number of environmental effects ...because of the lack of market price data or easily accessible direct proxy price data.' (OECD, 1992:49)

Given this neglect the OECD report (1986) on the public management of forests, must be regarded as a landmark publication amongst the reports reviewed in this study.

Overall, given that all economic activity centres on the idea of 'development', the general lack of incorporation of environmental issues is a serious omission. As a result the weaknesses identified in Australian assessment processes are not unusual when compared with 'world best practice'-though this does not make such deficiencies acceptable according to economic theory. Neo-classical economic theory would state quite clearly that on a worldwide basis, as well as locally, such an omission must be regarded as a threat to the economic well-being of society.

Existing Assessment Frameworks

Australia, with its federal political structure, has many different means by which choices over values are made. Rather than describe this diverse array of assessment frameworks, this section will provide a thumbnail sketch of the key features of assessment frameworks and processes. The review of various reports conducted in the previous section, along with the international examples cited, seem to point to three key characteristics of concern in any assessment:

Who Makes the Choices?

'Who makes the decisions' is significant because it determines whose values are brought to the process. The economics profession has long argued that institutions tend to be captured by the values of specific groups, and that to expect any public institution to be value-free and working purely in the public interest is naive. From the perspective of assessment processes, it is important that advocates of differing values are given an equal voice in the process of analysis and decision-making.

What Value Information do they have?

The RAC Forest and Timber Inquiry report took some interest in the Victorian model which involved the Land Conservation Council in decisions about selection of forest logging coupes. The AHC-CALM model of joint survey work also received some favourable attention, though its joint deliberations were designed to identify values rather than make decisions. That no other systems were singled out for favourable comment, in the various reports reviewed in this study, seems to indicate that other processes may be lacking an adequate contribution from representatives of other values.

It is not unfeasible that an appropriately structured institutional process could make optimising decisions-given that the right information on values was available. The information provided, as outlined by the thirteen reports reviewed in this process, is inadequate for the task of making choices over values. The AHC-CALM model received some favourable comments for its process from the ESD and RAC reports Certainly, its assessment method is systematic and thorough. Its prime weakness is that it does not consider, in its own words, 'social values' (AHC, 1992:29). 'Social values' as defined by the AHC, appear to cover that group of values described by economists as existence value. As a result, it only acknowledges some of the values considered important in a total economic assessment. That it was commended as a model for others suggests that even analysis of 'non-social' values is deficient in existing assessment frameworks and processes.

Other reports such as the Victorian Auditor-General's report (AG, 1993) and the NSW Public Accounts Committee (PAC, 1990) confirm the impression that only poor data is available on values.

Financial values are also a significant part of the analysis. A significant range of reports located deficiencies or made recommendations that pointed to opportunities to improve the gathering of financial data. The deficiencies in this regard extended to the data on both financial values generated by forestry and financial values generated by other activities. The existing assessment framework and processes were not being supplied with the relevant data because the necessary mechanisms such as appropriate accounting systems are not adequate (eg. RAC, 1992:429).

How do they Compare Different Values?

Assessment frameworks must make choices between values. Few technical models were offered on how such choices should be made,outside the political process. The Victorian Auditor-General's report noted that the previous system in that State had used a committee of relevant officials, backed up by a requirement to find options, with a final 'safety valve' being provided by there being recourse to the Minister. No doubt similar methods are used elsewhere, though, what is required is a method that considers all values at the point of decision with appropriate systemic checks and balances.

The AHC-CALM model and the process offered by the RAC Forest and Timber inquiry were the closest Australian examples of technical approaches which are beginning to seek full consideration of all values. The OECD (1986) report is a benchmark in this field. Only the process implied in the RAC report comes close to this analysis. It is assumed that if there were another system, the RAC inquiry would have located it. The existing assessment frameworks, obviously, can be said to rely on the judgements of those who take the decisions and the information that they have been supplied with. There must be serious doubts that these existing frameworks have been properly structured. A assessment framework and process needs to specify who decides, with what value information and by what intervalue comparative techniques.

Weaknesses in the Existing Assessment Framework

Inadequacies in Pricing/financial Valuation

The significance of non-financial values to the calculation of financial values is considerable. The forest agencies and all the reports reviewed in this study have adopted a compartmentalised approach to pricing. The compartmentalised approach assumes that assessment of choices between values occurred prior to the pricing of timber. If non-financial values are improperly assessed, then the resulting effect is that financial values will also be inappropriately founded. In this sense a compartmentalised approach to pricing is a practical manner of settling commercial pricing issues. In simple terms, if non-financial values are being compromised it is due to the lack of adequate assessment prior to the commercial pricing process.

Pricing of logs should, however, be offered on a normal commercial basis. It is fairly clear from the reports that have been reviewed that operations are not presently proceeding on a commercial basis. Prices of forest timber are being distorted by a range of factors that are, in general terms providing a subsidy to the logging of native timber, whilst other factors are raising the costs of private plantations (see NPAC, 1991). Other factors such as the lack of rational markets for the pricing of water (see Read and Sturgess, 1992) and other services, partly or wholly derived from forests, are raising the relative cost of alternative financial values.

The following distortions in commercial pricing of timber have been identified. The list has been 'cobbled' together from all the reports reviewed in this survey.

1. The lack of marginal cost pricing.
2. The non-existence of effective commercial accounting systems.
3. Long-distance transport subsidies.
4. Non-payment of rental for use of public land.
5. Low royalties; ie. low log prices for timber (see note below)
6. Possible cross-subsidisation from community service obligations.
7. No liability for payment of a resource rental tax.
8. No liability for payment of a notional income tax (presumably meaning company tax).
9. No liability for local government rates.
10. No liability for sales tax on vehicles, plant and equipment.
11. Non, or low, payments of interest on borrowed capital.
12. Non-payment of dividends.
13. Low user charges for cost recovery from long-distance transport
14. The lack of commercial discipline due to lack of a corporatised commercial structure;
15. Serious deficiencies in the tax system for plantations;
16. Problems with planning, rating and exporting from plantations;
17. Problems with the small numbers of buyers and sellers of timber;

18. Problems with the purchaser behaviour of large buyers of timber;
19. The dominant market role of state forestry agencies;
20. Failure to properly account for non-financial values in pre-logging operations either by exclusion of areas or by forestry codes of practice.
21. Failure to adequately assess the potential financial contribution from sales of goods and services such as water from the native forest.
22. Failure to separate commercial operations from regulatory functions.

They also felt that the market price was too low and that international prices were not as significant in pricing native forest timber as was considered by the Commission. (PAC, 1990:77) They felt that the native forest timber price should be raised to 'the point of indifference.' (?)(PAC, 1990:77) and that the case for doing so was 'overwhelming' (PAC, 1990,79). This is even more significant when it is considered that 'the Forestry Commission sets these prices...the Commission is clearly a price leader.' (PAC, 1990:97)

Inadequacies in Assessment Frameworks

The existing assessment framework is in a period of transition. The post-1945 charter of the forest agencies to push for economic development has changed. The capital resources of government and the Australian economy are scarce and need to be deployed for maximum financial benefit or for some clearly identified social 'good'. Changes in social values are also placing pressure on the old system. These two forces, in combination, are leading to greater scrutiny of forest activities, and demanding greater levels of accountability. It is clear that assessment processes are lagging in their response to these issues.

Entrepreneurial Capture

The existing assessment process is vulnerable to 'capture' by entrepreneurs proposing one-o ff developments. This limits government ability to consider a range of project options as proposed by the OECD (1986) and the World Bank (1992) in particular. The existing Australian approach of having a single project option placed in front of decision-makers must be regarded as inferior, if not a structural distortion of the decision-making process.

Lack of Commercially based Accounting of Projects

The lack of commercially-based accounting data for project analysis means that it is not possible to accurately evaluate true project financial worth. This means that many unprofitable activities are slipping through the policy net adding to the economic burden already being borne by Australians. The effect of such methods is to inflate the relative worth of financial values vis a vis non-financial values. It also inflates the relative values of forestry operations vis a vis the alternative activities such as plantations and water production.

Lack of a Comprehensive set of Values Descriptions

The next major flaw in assessment processes is the lack of a complete set of values relating to forestry. Each report examined has mentioned a range of words that seek to describe, in particular, the non-financial values, but there has not been a report which systematically listed what society derives from the forest. It is little wonder then that assessment frameworks barely exist or function if the values are not even identified.

Lack of Focus on the Full Suite of Values

Where non-financial values are considered, they tend to relate only to the assessment of physical characteristics of the forests. All reports acknowledge the need to protect these physical characteristics, for example 'sustainable yield', yet there is evidence that they are not being protected, even at this rudimentary level ((PAC, 1990:55). If physical characteristics are only barely being protected, then it is reasonable to assume that other more abstract and intangible values of the forest are not being taken into account. The Dorrigo Management Area Environmental Impact Statement (FCNSW, 1992:i) does not mention any non-financial values other than soil, birds, animals and recreation. Ecological sustainability and existence values are barely considered at all.

Public Involvement: Preached by Some but not Practised

The practice of assessment seems to place too little stress on the need to obtain public review and input into forestry decisions. In an area where clashes over values are common, the need for representation by different values should be considered. As a result the political process is starved of information about social preferences. The OECD (1986) reveals a considerable commitment to public involvement, in particular their example of the 'Inner Valley Road Project'(OECD, 1986:34).

NFV's are not being Assessed

The most serious flaw is that it appears that assessment of non-financial values is simply not occurring. The concentration of reports on discussing aspects of how such analysis could be done, rather than a focus on how it is being done leads to the conclusion that serious analysis of the relative worth of non-financial values is yet to begin. It would seem fair comment that if relatively simple choices between wood and water production are not being evaluated (eg. Read and Sturgess, 1992), then other more complex interactions are also being neglected.

Future Generations Considered by Implication Only

The OECD (1986) report was not without its flaws. Its assessment framework gave little attention to considering the impact of future generations from any valuation assessment process. In economic theory it is impossible to internalise the interests of future generations simply by assessing the preferences of the present generation. Neither market processes or valuation techniques, in the face of irreversibility, are capable of 'looking after' the interest of future generations in any manner that can be relied upon. Fortunately the concept of ecological sustainability goes some way to achieving this goal. The internalisation of the interests of future generations requires explicit discussion.

Most reports steered clear of identifying procedures that would resolve the choices thrown up in the forestry debate. The RAC report (1992) went furthest in this regard but tended to educate by demonstration rather than explicit prescription. The ESD report (1991) had some brief comments, but if the assessment of these different values is to proceed satisfactorily, it is clear that the hard issue of 'assessment frameworks' must be dealt with honestly.

The OECD (1986 and 1992) indicated that other nations are moving towards direct evaluations of non-financial values. Australia's approach would appear to be somewhat slower than average, though of course there are some institutions such as the World Bank that have acknowledged the need for change but are yet to make the practical changes that are required. It was significant that the United States was acknowledged as having greater experience in the area of non-monetary evaluation. The experience was deemed to be due to the 1969 Environmental Policy Act which required quantification in non-monetary terms of all environmental impacts. The OECD concluded that it would seem that education and more valuation projects are a prerequisite for more effective utilisation of the techniques. A learning process is needed. (OECD, 1992:38).

Valuation techniques, whether monetary or non-monetary, form the backbone of an assessment framework. The same learning process is required for assessment processes as well. Many OECD countries are learning but Australia does not appear to have started.

Assessing the Assessment Frameworks

A major difficulty with evaluating an assessment framework is the sheer scale of the task itself, as anyone who has read an Environmental Impact Statement will testify. The result is a tendency to focus on one or two key angles to the exclusion of other angles. Table 6 seeks to provide a systematic 'framework' for assessing assessment frameworks. A series of questions is provided, that an assessment framework, if it is to be successful, should be able to address. The existing Australian and international assessment practices are evaluated The table has been tested on actual reports, in this case the Victorian Auditor General's report (AG, 1993) as well as an analysis of the Dorrigo Management Area Environmental Impact Statement (FCNSW, 1992).

The native forest is capable of producing three major categories of values. These are:

- financial values from timber production;
- alternative financial values from other forms of forest use; and
- non-financial values.

  This report has sought to examine the issue of giving each type of value 'full and due consideration' in policy processes. Three themes have recurred throughout this study.

- There is a theme focusing on the analysis of technical comparisons between different values. That is the use and abuse of various techniques that quantify values.

The native forest is capable of producing three major categories of values. These are:

- financial values from timber production;
- alternative financial values from other forms of forest use; and
- non-financial values.

  This report has sought to examine the issue of giving each type of value 'full and due consideration' in policy processes. Three themes have recurred throughout this study.

- There is a theme focusing on the analysis of technical comparisons between different values. That is the use and abuse of various techniques that quantify values.
- Asecond theme revolves around the nature of the institutional arrangements that ensure value selection which is socially optimal.
- The major theme, however, is achieving a level playing field between the major types of values derived from the forest.

The emphasis on achieving a level playing field is in the spirit of the microeconomic reform process that has taken place over recent years, and continues with recent reports such as the Hilmer Inquiry (NCP, 1993). Sims described microeconomic reform as occurring when changes are made to achieve more output from a given level of inputs. (Galligan et al, 1993:15).

The output from the forests is intended to maximise the social welfare of the community. For this reason, microeconomic reform in the forest sector must seek to arrange the three sets of values above so that welfare is maximised.

The implication of the need for micro-economic reform is that the sector is not making an optimal contribution to the quality of Australian life. This review of reports on the forest sector suggest that there is considerable potential for microeconomic reform in two dimensions.

- In broad social terms it would appear that the community is not maximising the social benefit from the forest resource.
- Secondly, within the social benefit the community receives, the financial value to be had from the forest resource is not being maximised, assuming that all other types of values remain unchanged (ceteris paribus 'All other things being equal.').

If the microeconomic reform process (optimising social benefits) in the forest sector is to succeed, the estimation of the worth of the three categories of value described above must be accurate. It would appear that estimation of these three categories of value is presently inaccurate. This inaccuracy would confound any attempt to derive optimal social benefits.

Effects of Miscalculating Financial Values

The impact of miscalculating the financial value of timber production is significant for Australia's macro-economic performance. Firstly, it overestimates the financial worth of timber production vis a vis some other form of commercial activity. The Read Sturgess report (1992) made it clear that it was more financially valuable for the Upper Thomson catchment to supply water than to supply timber, and consequently more profitable for the economy. The overestimation of the worth of forestry would tend to hide this potential economic 'free lunch'. Other competing uses for the forest may be similarly affected. More generally, the over-valuation of forestry financial values disguises what may be net losses in this sector which are actually diminishing the wealth of Australia unlike other commercially viable industries.

Failure to accurately value forestry also leaves the commercial operations of the forestry commissions vulnerable to an income shortfall in the event of corporatisation. Even without corporatisation, the poor measurement processes for forestry blind the commissions to other profit centres such as leasing forests to supply water to regional communities with expanding industries.

Over-valuation of forestry through poor input pricing, in addition to affecting competing uses of the forest, also undermines industries that provide substitutes for forest timber ranging from plantations to steel production. Inefficient input pricing in native timber production undermines all competing industries, all other things being equal. This report has focused on the 'collateral damage' caused to the plantation industry. However, what applies to the private plantation industry, also applies, probably on a lesser scale, to other industries that compete with native forest timber. The private plantation industry (NPAC, 1991) is one clearly visible victim of an 'unlevel' playing field in the forest sector.

The over-valuation of native forest harvesting undervalues, by definition, the product of plantations. This leads to lower levels of investment, employment and wealth creation than may otherwise have occurred. It prevents the development of value-added timber industries that can further develop on the basis of a plantation resource. This presents the Australian economy with a 'double loss'. The macro-economy loses, marginally at least, from harvesting native forests, and then loses again through the disincentive effects on plantations. Other industries may suffer as well. Finally, there is a third loss built into the playing field, which are the microeconomic distortions, such as poor taxation arrangements, identified by NPAC (1991), that inhibit the plantation industry. The microeconomic distortions in the plantation sector alter the financial calculus such that the value of native timber is enhanced relative to plantation timber. In addition these microeconomic distortions cause a further round of macro-economic losses from the plantation industry.

The structural distortions within the forestry sector, and within other sectors both competing and substitute, compound the overvaluation of forest timber relative to other values. This causes a loss of income, jobs and a waste of capital as private agents respond to these inappropriate signals.

The Effects of Miscalculating NFV's

The second major impact of overvaluing forestry financial values, is felt on non-financial values. Over-estimating of forestry financial values also underestimates, by definition, the relative worth of non-financial values. Thus the process of evaluating non-financial values is seriously handicapped from the beginning. The significance of this was pointed out by Sims, in another context, when he stated

> Clearly our objectives are broader than maximising GDP. For example pollution in our cities is of general concern. It might be that including externalities in prices would decrease environmental damage and decrease growth, but improve overall living standards. (Galligan et al, 1993:17, emphasis added)

Two points flow from the comments by Sims.

Firstly, Sims appears to be giving emphasis to a commonly-held view that environmental improvement can only be achieved at a cost to the economy. This review has indicated, in a very clear manner, that in the forest sector financial gains (see note below) can be achieved whilst making environmental improvements. The view espoused by Sims presumes that the 'balance' between environmental and financial considerations is 'efficient'; that is, in the context of this report, the evaluation of the financial worth of native forests has been accurately calculated. The evidence indicates that the balance is inefficient, because calculations of financial and alternative financial values, as well as NFV's are inaccurate. Therefore, this report indicates that improvement in environmental conditions, over a very large margin, will improve financial conditions as well.

At some point along the policy continuum, improving environmental conditions will negatively affect financial conditions-but the forest sector, along with many other sectors, is not at this point, and indeed is a long way from it. Microeconomic reform in the forest sector has the potential to bring environmental improvement in its wake by removing financial inefficiencies that have contributed to over-exploitation of the environment.

The implication of financial 'inefficiency' in the sector needs to be clearly understood. It implies that society could be financially better off if certain microeconomic reforms were undertaken. The implication of 'economic' inefficiency is that, in the context of this report, the environment and the financial wealth of society, could be better off, leading to an unambiguous improvement in society's level of 'satisfaction'. In this sense, when 'financial inefficiency' is present, microeconomic reform and environmental policy can be complementary, and not competing policies.

Secondly, failure to evaluate non-financial values relative to financial values accurately is critically important to maximising 'overall living standards'. The benefits that are provided to Australians from NFV's can also be described as 'psychic income' (see note below).

These benefits also have the fortuitous 'collateral' benefit of also attracting tourists, and hence contributing to the expansion of this critical export industry.

Non-financial values provide direct inputs into the daily consumption of Australian citizens and even to foreigners through tourism, wildlife films and existence values. An assessment framework which seeks to optimise benefits to the community, but which does not recognise these non-financial values can generate financial flows that actually undermine community well-being.

This happens because the source of much 'psychic income' (ie. the natural environment) has been lost and replaced by financial benefits that are too small to compensate. If non-financial values are to be given their full and due consideration, then financial and alternative financial values must be correctly evaluated.

This review has identified that there are significant inefficiencies in the estimation of value in the forest sector. The presence of these inefficiencies in the market activities of the forest sector is due to structural distortions.

The structural distortions contribute to poor financial and environmental performance. There is a considerable social, environmental and financial dividend to be achieved by microeconomic reform in the forest sector. Microeconomic reform requires that a 'trilogy' of assessments be made: evaluate financial values (FV's), evaluate alternative financial values (AFV's) and evaluate non-financial values (NFV 's). Continuation of present practices such as failure to 'estimate value' accurately will cost Australia in lost jobs, lower incomes and a diminished environmental quality of life.

Here this report distinguishes between financial gains that contribute to the overall health of the financial economy, and economic gains that contribute to the overall state of social well-being. Improvements in financial well-being are commonly referred to as economic gains, for example improved resource allocation, because of the easy but methodologically sloppy habit of referring to the financial activities of the Australian society as the 'economy'. Accurately the economy is the process by which we gain 'welfare', a part of which is the financial gains offered by financial activity.

In another circumstance, when a Treasury officer was asked to described the non-pecuniary benefits of working in Treasury, he called these benefits 'psychic income'.

THE ECONOMIC DIMENSION OF TRANSITION

One way in which an economist familiar with cost-benefit analysis (CBA) may look at environmental impact assessment (EIA) is that EIA is a CBA requir-ing a large amount of information rigorously collect-ed and analyzed by a group of scientists of various disciplines. One can appreciate that the development of EIA has served to advance the systematic descrip-tion and quantification of environmental effects in a way which can only serve to improve the quality of CBA undertaken.

One can also appreciate that EIA constitutes a process, formalized through environ-mental legislation, regulations and procedures, which in turn reinforces the community participation re-quired for informed decision-makers seeking the common good. While CBA within EIA is a key gen-erator of information to the decision-makers, recog-nition should also be given that the EIA process is both a planning tool and an execution monitoring tool. Thus, the role of CBA is not as an intermediate step, but rather as an integrated parallel process.

Hence, the CBA should follow up, quantify and eval-uate the dynamic iterations created by the exhaustive search of alternatives, by the efficient inclusion of en-vironmental protection and mitigation measures and by the explicit valuation of preference and choice ar-ticulated by community, society and their representa-tives acting in the best interest of present and future generations.

VALUE OF CONSIDERING ENVIRONMENTAL EFFECTS EARLY ON IN THE PROJECT CYCLE

The main purpose of the economic analysis of a project is to ascertain whether the project can be ex-pected to create more net benefits than any other, mutually exclusive option, including the option of not doing it. Consideration of alternative options, therefore, is a key feature in proper project analysis. Often, important choices about alternative project options are made early on in the project cycle.

These options may differ considerably in their general eco-nomic contribution, and they may also differ greatly concerning their environmental impact. Therefore, including environmental effects in the early econom-ic analyses, however approximately, should improve the quality of future decisionmaking.

The CBA process-defining objectives, searching for alternatives, costing out the resources involved, speci-fying the effects of each option involved and compar-ing all the costs and benefits-normally requires considerable efforts. In the case of Cuba, we are faced with two joint families of economic valuation prob-lems: the "environmental" ones and those problems associated with the incipient transition from a cen-trally planned economy to a truly participative mar-ket based economic system.

In principle, economic analyses are to take into ac-count all costs and benefits of a project. With regard to environmental impacts, however, there have been two basic problems even in developed countries.

First, environmental impacts are often difficult to measure in physical terms.

Second, even when im-pacts can be measured in physical terms, valuation in monetary terms can be difficult. In spite of such dif-ficulties, a greater effort needs to be made everywhere to 'internalizes' as many environmental costs and benefits as possible by measuring them in money terms and integrating these values in the economic appraisal.

The measurement in money terms are made even more difficult in countries undergoing market reforms.

The environmental valuation problems can be sum-marized as follows:

VALUE OF ENVIRONMENTAL ASSETS

While man-made and human capital may be valued with relative ease by observing existing market sys-tems, where available; the existence value of clean wa-ter and air, tropical forests, wetlands, coral reefs and other environmental assets and their functions is much more difficult since not even market prices can reflect their full contribution to other economic ac-tivity and to human welfare.

In particular, the mar-ket price of water does not reflect the various services it provides nor do market values can accurately re-flect what happens when irreversible loss or damage of natural resources occurs as environmental degrada-tion exceeds a critical threshold level.

Complexity of environmental valuation also arises due to the multi-ple functions of being a source of raw materials and energy, being a sink for assimilating man-made wastes and providing other services such as recre-ational/tourism services, storage of genetic diversity and scientific and educational benefits.

The following classification is useful:

- Direct use values are derived from the economic uses made of the natural system's resources and services. Examples of these are outputs such as timber, game and recreation from forests or fish and scuba tourism from coral reefs.
- Indirect use values are the indirect support and protection provided to economic activities and property by the resource system's natural func-tions or environmental services. Examples of these are watershed protection and soil erosion prevention provided by forests, and beach sand and mooring facilities protection provided by coral reefs.
- Non-use values lie in the special attributes of the natural system as a whole, its cultural and heri-tage uniqueness; it includes both existence and option values. Existence values reflect public goods which can be enjoyed by more than one consumer without decreasing the amounts en-joyed by others such as clean air, beaches and forests. The existence value is the utility that consumers derive from just knowing the public good exists. A way of measuring that utility would be to measure the willingness to pay or the contingent value assigned if the public good were to disappear. Option values reflect what current generations wish to bequeath for future generations to inherit. They imply both an ethi-cal commitment to sustainabilty for the children of our children and in a shorter time-frame the maintenance of options to solve current prob-lems. Examples of option values have been devel-oped for forests relating to biodiversity and the search for cures of cancer and AIDS-"if forest were to disappear then the options to find such cures will vanish" or "as long as the forest is pro-tected there is the option to find the cure".

Direct Effects Valued on Conventional Markets Some methods are directly based on market prices or productivity. This is possible where a change in envi-ronmental quality affects actual production or pro-ductive capability.

Change-in-Productivity: Development projects can affect production and productivity positively or neg-atively. The incremental output can be valued by us-ing standard economic prices where available. Loss-of-Earnings. Environmental impacts can signifi-cantly affect human health. In theory, the value of health impacts should be determined by the willing-ness to pay of individuals to maintain their health. In practice, one uses earnings lost upon early death, dis-ease or job absence. This approach is used in highway and industrial safety, and in air pollution studies.

The "implicit value of human life" approach is reject-ed by many as dehumanizing since human life can be said to have infinite value. However, society, govern-ment regulations, insurance companies and judicial courts implicitly and explicitly place finite values on human life and health. This is a necessity reflecting limited resources to be allocated for health expendi-tures. The relatively high level of health expenditures in Cuba would indicate a high implicit value of hu-man life and health. However, one can also discern a political motivation and its attendant benefits behind it.

Preventive Expenditures: Individuals and governments invest in prevention measures to avoid or reduce un-wanted environmental effects. Environmental dam-ages, are often difficult to assess, but historical infor-mation on preventive measures and their costs may be interpreted as a minimum value for the expected benefits that the preventive measures seek. If it is found, for example, that there is a historical pattern of under investment in Cuba for natural disaster planning and prevention; then, it can be inferred that the benefits expected from such preventive measures have been very low. This conclusion would in turn imply that the loss of human life would be assigned a relatively low value. While this possible conclusion may appear to conflict with the high political priority for health expenditures, it can be observed that the level of damages or loss of lives caused by natural di-sasters cannot be easily attributed to government in-action and thus the political cost may be easily ex-plained away.

Potential Expenditure

Valued on Conventional Markets

Replacement Cost: Simply, the costs that would have to be incurred in order to replace a damaged asset. The estimate is not a measure of benefit of avoiding damage since the damage costs may be higher or low-er than the replacement cost.

Valuation Using Implicit (or Surrogate) Markets

Sometimes one must use market information indi-rectly. Approaches to be considered are the travel cost method, the property value approach, the wage dif-ferential approach, and uses of marketed goods as surrogates for non-marketed goods. Each technique has its particular advantages and disadvantages, as well as requirements for data and resources. One must determine which techniques might be applica-ble to a particular situation.

Travel Cost: This approach measures the travel cost (and time involved) which reflects the willingness of consumers or users can serve to measure the benefits produced by recreation sites (parks, lakes, forests, wilderness). It can also be used to value "travel time" in projects dealing with fuelwood and water collec-tion.

Property Value: This valuation method is based on the general land value approach and can determine the implicit prices of certain land areas. The property value approach can help analyze willingness to pay for properties with different pollution levels and infer the implicit cost of pollution. The method compares prices of houses in affected areas with equal size and similar neighborhood characteristics elsewhere in the same metropolitan area.

Valuation Using Constructed Markets

Contingent Valuation: When society's preferences as revealed in market prices are not available, the con-tingent valuation method tries to obtain information on individual preferences by posing direct questions about willingness to pay. It basically asks people what they are willing to pay for a benefit, and/or what they are willing to accept by way of alternate compensa-tion to tolerate an environmental cost. This process may be achieved through a direct questionnaire/sur-vey. Willingness to pay is difficult to measure and de-pends on the income level of the sample subjects, and involves problems of designing, implementing and interpreting questionnaires. While its applicability may be limited, there is now considerable experience in evaluating the quality of supply of potable water and electricity services.

Artificial Market: Such markets can be constructed for experimental purposes, to determine consumer willingness to pay for a good or service. For example, a home water purification kit might be marketed at various price levels, or access to a game reserve may be offered on the basis of different admission fees, thereby facilitating the estimation of values placed by individuals on water purity or on recreation facilities.

THE DISCOUNT RATE

Discounting is the process by which costs and bene-fits occurring in different time periods may be com-pared. The discount rate to be used has been a gener-al problem in cost-benefit analysis, but it is particularly important with regard to environmental issues, since some of the associated costs and benefits are very long-term or irreversible in nature. In stan-dard analysis, past costs and benefits are treated as "sunk" and are ignored in decisions about the present and future. Future costs and benefits are discounted to their equivalent present value and then compared. In theory, in a perfect market, the interest rate re-flects both the subjective rate of time preference (of private individuals) and the rate of productivity of capital.

Higher discount rates may discriminate against fu-ture generations. This is because projects with social costs occurring in the long term and net social bene-fits occurring in the near term, will be favored by higher discount rates. It is often argued that discount rates should be low-ered to reflect long-term environmental concerns and issues of intergenerational equity. However, this would have the drawback that not only would eco-logically sound activities pass the cost-benefit test more frequently, but also a larger number of projects would generally pass the test and the resulting in-crease in investment would lead to additional envi-ronmental stress.

Many environmentalists believe that a zero discount rate should be employed to protect future genera-tions. However, employing a zero discount rate is in-equitable, since it would imply a policy of total cur-rent sacrifice, which runs counter to the proposed aim of eliminating discrimination between time periods-especially when the present contains wide-spread poverty.

In the case of projects leading to irreversible damage (e.g., destruction of natural habitats, etc.), the bene-fits of preservation may be incorporated into stan-dard cost-benefit methodology using the Krutilla-Fisher approach. Benefits of preservation will grow over time as the supply of scarce environmental re-sources decreases, demand (fueled by population growth) increases, and possibly, existence value in-creases. The Krutilla-Fisher approach incorporates these increasing benefits of preservation by including preservation benefits foregone within project costs. The benefits are shown to increase through time by the use of a rate of annual growth. While this ap-proach has the same effect on the overall CBA as low-ering discount rates, it avoids the problem of distort-ed resource allocation caused by arbitrarily manipulating discount rates.

CONCLUSIONS

In order to achieve economically sustainable manage-ment of natural resources and environmental protec-tion, one must effectively incorporate environmental concerns into decision making through the EIA pro-cess.

This presentation has reviewed concepts and tech-niques for economic valuation of environmental im-pacts within EIA procedures. The process of internal-izing these environmental externalities can be facilitated by making rough qualitative assessments early on in the project evaluation cycle-the advan-tages of which would include:

- early exclusion of alternatives that are not sound from an environmental point of view;

- more effective in-depth consideration of those al-ternatives that are preferable from the environ-mental viewpoint; and
- better opportunities for redesigning projects and policies to achieve sustainable development goals.

In order to fully reflect society's values and preferenc-es about environmental values, non-market methods of estimation can be used and will enhance commu-nity participation through well designed and admin-istered questionnaires and surveys. Research and training about EIA and embedded economic analysis methods is needed in Cuba. As developing countries learn to successfully apply EIA methodologies the goals of sustainable development will become more attainable.

Bibliography

Aukland, L.: (1997). *The Effect of Selective Logging on Amphibian Diversity of Budongo Forest*, Uganda. B. A., Oxford University.

Babweteera, F.: (1998). *Influence of Gap-size and Age on the Diversity and Abundance of Climbers in Budongo Forest Reserve,* Uganda. M. Sc., Makerere University.

Bahati, J.: (1995). *Impact of Arboricidal Treatments on the Natural Regeneration of Species Composition in Budongo Forest,* Uganda. M. Sc., Makerere University.

Barrows, M.: (1996). *A Survey of the Intestinal Parasites of the Primates in Budongo Forest,* Uganda. Glasgow University.

Bennet, H.: (1996). *Study of Kinship, Social Relations and Possible Avoidance of Incest among Budongo Chimpanzees*. M. A., Edinburgh University.

Bethell, E.: (1998). *Vigilance in Foraging Chimpanzees*. University College, London University.

Fairgrieve, C.: (1995). *The Comparative Ecology of Blue Monkeys (Cercopithecus mitis stuhlmanni) in Logged and Unlogged Forest, Budongo Forest Reserve,* Uganda, Edinburgh University.

Hills, C.: (1998). *An Analysis of Phylogeny and Gene Flow in a Population of Pan troglodytes schweinfuthii*. M. Sc., Oxford University.

Johnson, K.: (1993). *Local Uses of Budongo's Forest Products*. M. Sc., Oxford University.

Johnson, K.: (1996). *Hunting in the Budongo Forest,* Uganda. SWARA.

Johnson, K.: (1996). *Local attitudes towards the Budongo Forest,* Western Uganda. Indigenous Knowledge & Development Monitor.

Miller, P. L.: (1993). *Some dragonflies of the Budongo Forest,* Western Uganda. Opusc.

Mork, K. & Solvang, R.: (1996). *Edge Effects on an Understorey Bird Community in Budongo Forest Reserve*, Uganda, Agricultural University of Norway.

Musamali, P.: (1996). *The Ecology, Diversity, and Relative Abundance of Small Mammals in Primary and Disturbed Forest Compartments in the Budongo Forest Reserve.* M. Sc., Makerere University.

Owiunji, I.: (1996). *The Long Term Effects of Forest Management on the Bird Community of Budongo Forest Reserve*, Uganda. M. Sc., Makerere University.

Reynolds, V.: (1965). *Budongo: A Forest and its Chimpanzees*, London, Methuen.

Reynolds, V. & Reynolds, F.: (1965). *Chimpanzees of the Budongo Forest*, New York, Rinehart and Winston.

Shiel, D.: (1997). *The Ecology of Long-term Change in a Ugandan Rainforest*. D. Phil., Oxford University.

Spini, L.: (1998). *The Frequency and Significance of Social Grooming in the Sonso Community of Chimpanzees at the Budongo Forest*, Uganda. M. Sc., Oxford University.

Index

A

B

C

□□□